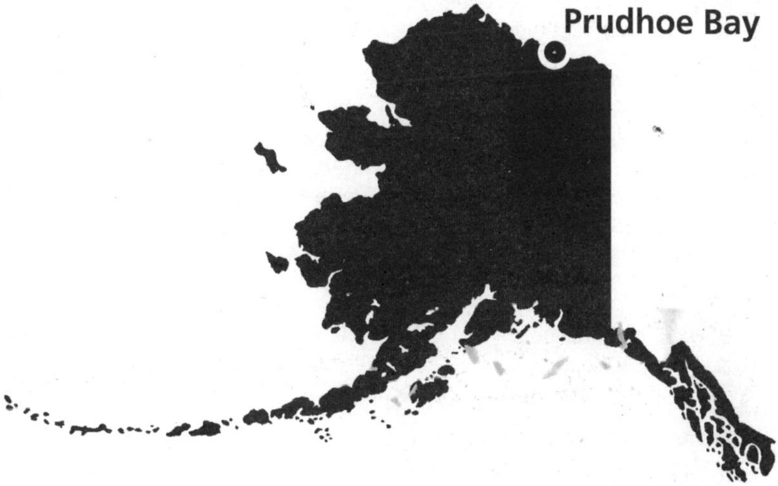

Discovery at Prudhoe Bay

by
John M. Sweet

ISBN-10 0-88839-630-9
ISBN-13 978-0-88839-630-3

Cataloguing in Publication Data

Sweet, John M. (John McCamey), 1924–
Discovery at Prudhoe Bay / by John M. Sweet.

ISBN 978-0-88839-630-3

1. Petroleum—Prospecting—Alaska—Prudhoe Bay Region—History. 2. Petroleum industry and trade—Alaska—Prudhoe Bay Region—History. 3. Atlantic Richfield Co.—History. 4. Trans-Alaska Pipeline (Alaska)—History. 5. Sweet, John M. (John McCamey), 1924–. 6. Prudhoe Bay Region (Alaska)—History. 7. Petroleum geologists—Alaska—Biography. I. Title.

HD9567.A4S94 2007 338.2'7282097987 C2007-906961-4

Printed in Indonesia — TK Printing

Editors: Nancy Miller, Theresa Laviolette

Production: Mia Hancock

Cover Design: Mia Hancock

Front cover images: John Sweet

Back cover photographer: Don Spear

Figure illustrations: Jack Nason

Published simultaneously in Canada and the United States by

hancock house

HANCOCK HOUSE PUBLISHERS LTD.
19313 Zero Avenue, Surrey, B.C. Canada V3S 9R9
(604) 538-1114 Fax (604) 538-2262

HANCOCK HOUSE PUBLISHERS
1431 Harrison Avenue, Blaine, WA U.S.A. 98230-5005
(604) 538-1114 Fax (604) 538-2262

Website: **www.hancockhouse.com**
Email: **sales@hancockhouse.com**

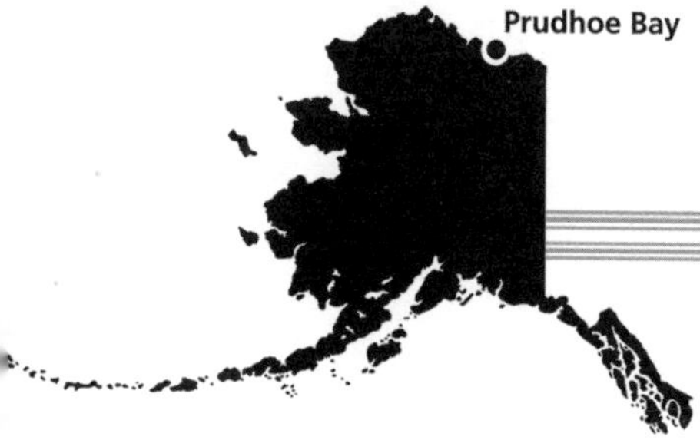

Contents

Dedication

This tale could not have been told without the participation of the dozens of individuals who played a part in its unfolding over a period of nearly six decades. I dedicate it to the memories of those who have passed on, and in honor of those who remain.

In memory of...

Charles Selman, Alaska district geophysicist, playing in a bridge game at a mobile seismic camp surveying for Richfield Oil Company in December 1963 with a temperature outside of –8°F. Photo by Gar Passel.

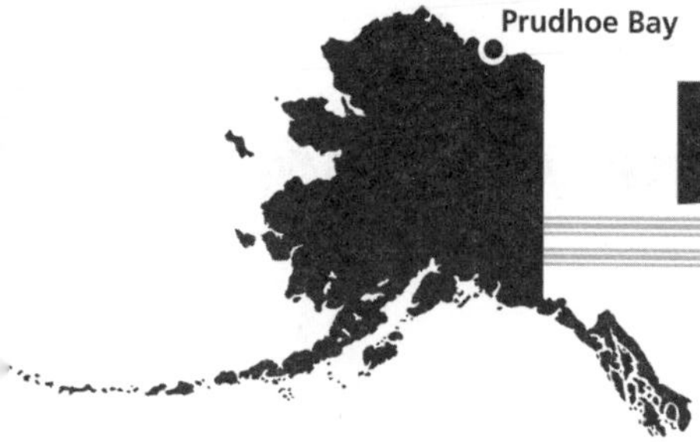

Foreword

This is the story of one of the greatest adventures of the twentieth century. Few books are being written about it, because the heroes aren't poets. They are pioneers who understand rocks and reservoirs, fossils and flow rates. They had to convince upper management to send them to the high Arctic to find oil, and they had to invent ways to work and survive at seventy degrees below zero. They are rugged and talented men, and John Sweet is one of the best. Fortunately, he has turned his love for a challenge and his admiration for his fellow believers into valuable history. It's an engrossing tale.

Their crusade was scoffed at by their competitors. Not only did they have to find oil, they had to find it in quantities never before discovered in America — billions of barrels of it — or it would never be pumped out of the ground and shipped to market. To most experts, their gamble seemed ridiculous.

As the second governor of the young state of Alaska, I disagreed. I saw the vision and shocked the industry, and even the White House, by predicting that 40 billion barrels of precious oil lay under the two-mile-thick permafrost blanket that stretches across Alaska's North Slope. John and his colleagues persevered, and proved me right. As a result, Alaska has never been the same, and we are proud of their stunning accomplishments.

Walter J. Hickel
November 18, 2004

Acknowledgments

Many of those involved in the long line of events leading to the Prudhoe Bay field discovery have been most gracious in helping me with this project. They have been so helpful I want to give credit to all of them, but the danger is that I will miss someone. My special thanks go to those who provided publications, letters, and drafts of interviews, personal letters, telephone calls, faxes, encouragement, and dozens of e-mail exchanges. The list is in alphabetical order and includes their former employers.

Ann Byrd Allen, ARCO; G. Ray Arnett, Richfield, Dept. of Interior; Julius Babisak, Atlantic, ARCO; Phyllis Eileen Banks, Anchorage Borough, friend; Edward M. "Mo" Benson, Richfield and ARCO; Lonnie Brantley, Atlantic, ARCO; Les Brockett, Richfield, ARCO; Allen Brown, Richfield, ARCO; Ed Capen, Atlantic, ARCO; Dick Cathriner, Richfield, ARCO; J.B. Coffman, Humble Oil, Exxon; Richard W. Crick, Atlantic, ARCO; John Etnyre, Sinclair, ARCO; George Gryc, USGS; Al Hakeem, ARCO; David Hite, PhD, Atlantic, ARCO; J.R. Jackson, Humble Oil, Exxon; Kathy Bouchor, ARCO; Bruce Campbell, State of Alaska; H.C. "Harry" Jamison, Richfield, ARCO; Thomas R. Marshall, Alaska Land Selection Officer; Dean L. Morgridge, Humble Oil, Exxon; Charles "Gil" Mull, Richfield, ARCO, Humble, AKDNR;

Gaël Mustapha, writer; Glenda and Milton Norton, Richfield, ARCO, William C. Penttila, Consultant, Atlantic, ARCO; Michael Pichter, ARCO; Colleen Rutledge, ARCO; Gene Rutledge, Author; Mickey Sexton, Sinclair, ARCO; Thelma Shear, ARCO wife; Jerry P. Siok, British Petroleum; Fred Sollers, Humble Oil, Exxon; Robert Specht, Richfield, ARCO; Arman Spielman, Richfield, ARCO; James Stover, Richfield, ARCO; Irvin Tailleur, USGS; Leland B. Wilson, Atlantic, ARCO.

Extra thanks goes to Julius Babisak, Harry Jamison and Bob Specht, from whom I obtained much valuable information from oral history interviews and professional reprints they provided.

Thanks to George Gryc and Irvin Tailleur for providing technical papers on Arctic Alaska geology, photographs and consultation.

My good friend and co-worker for many years, Dick Crick, obtained permission from Phillips Petroleum to search and copy pertinent documents. Dick spent hours on this invaluable assistance. Thanks, Dick.

Lee Wilson and Larry Bell contributed information on drilling techniques. Lee was always encouraging, rooting for this story to be told.

Dave Hite provided valuable insights to the cone delta depositional environment at the Prudhoe Bay Oil Field.

Barb Burch contributed photographs taken by her stepfather, Walton G. Banks, some of which appeared in the book *Silent River*.

Don Spear has a long attachment to photography and has become expert in digital enhancement of photos. He graciously enhanced many of the photos and figures.

Gaël Mustapha, a writer acquaintance, provided valuable suggestions and handholding.

Elizabeth O'Leary, the librarian in the Geoscience Library, University of Arizona, was kind and helpful with references and fulfilling e-mail letter requests.

Thanks to Clay S. Turner of Roswell, Georgia for providing us with a copy of his very beautiful parhelia ("sun dog") photograph.

Charles R. Metzger's *The Silent River* provided much information on the adventures of George Gryc's 1945 USGS geological field party, which he graciously gave me permission to use. Thanks Mr. Metzger.

Jack Roderick gave me much encouragement and counsel, and his book *Crude Dreams* is a very good and interesting history of the oil industry through the years. I used, with attribution, some of his tales. Thanks Jack.

Since meeting Marvin "Marv" D. Mangus in the late 1950s, he has been a lifelong friend, co-worker and a rich source of information from his tenure in Alaskan geology, which began with the NPR4 in 1946 as a USGS geologist. Jane and Marv have lived in Alaska since 1962. The Anchorage art community honored him in 2005 for his highly prized accomplishments as a landscape artist. Marv has shared much with me, many photos, many conversations and his section of history recorded in connection with an ARCO history project, some of which is included in this story. He is one of the Deans of northern Alaska geology and is as much a part of the Prudhoe story as anyone, and appears in this narrative frequently. Thanks Marv.

Thanks to Gil Mull for reviewing, correcting and offering suggestions on parts of the manuscript and sending me various publications, file material and many e-mail letters and, toward the end of the project, providing me with colored slides of many North Slope activities. These all helped immeasurably.

Phillips Petroleum bought ARCO properties in northern Alaska as a government condition of the merger of ARCO and British Petroleum. I thank the Phillips management in Anchorage for making it possible for Dick Crick to search their files.

Glenda and Milton Norton graciously shared several pages from their personal journal, and pictures from their slide files, about their earthquake experiences. Thanks Milt and Glenda.

Lael Morgan, longtime Alaskan reporter, writer, editor, publisher and teacher gave me invaluable advice and assistance in the rewriting of this work. Thanks, Lael; I was thrilled to have had your involvement.

Thanks to Mary Kay Grosvald, Karen Piqune, and others in the AAPG library, for responding to several pleas for help.

Phyllis Eileen Banks, a longtime friend and former Alaskan, edited, held my hand via e-mail and typed an early manuscript. When my frustration began to show, she always knew how to get me back on an even keel. I am grateful for her many skills, and most of all the long friendship with her and her husband, the Rev. Dr. Hal Banks.

Thanks to Gov. Walter J. Hickel, the second governor of Alaska and former Secretary of Interior in the Nixon Administration, for the Foreword to my book and for the quotations from his book *The Alaska Solution*.

Thanks also to Richard Klemer and Christi Hansen for reading an early manuscript and making valuable suggestions. Richard Wortman graciously volunteered to proofread the manuscript, to which he made many very valuable corrections and editorial recommendations. This heartwarming and professional help is greatly appreciated. Thanks, Rich.

I engaged and worked with Sandra H. Luber, a professional editor, for more than a year to guide me in improving the flow of the story and straightening my Pennsylvania Dutch syntax. It was a joy working with her and learning as we progressed. She became a friend and consultant.

I thank Meg Park for typing several preliminary and the final manuscripts. Meg is reliable, considerate, cooperative, prompt, efficient, pleasant and professional in every way. I never ceased to be impressed with her ability to handle any request, computer or otherwise.

Jack Nason is a wonderful golf partner, and a casual

conversation after a game led to him volunteering to work with the manuscript. Little did I know how deeply he would be become involved, and the great talent he has for design and implementation of computer-assisted publishing. Thanks, Jack.

Clare Flemming and Ryan Haley of The Explorers Club in New York combed their archives for several items about Ernest Leffingwell, a charter member. Thank you, Clare and Ryan.

Bruce Campbell with the State of Alaska kindly sent me a copy of *Thirty Summers and a Winter* by Evelyn Mertie, wife of one of the 1924 USGS geologists in the story. All efforts to secure permission to use material from her book came to dead ends.

Thanks to Dave Jackson, a workout friend and former weekly contributor of a column for seniors in the *Boulder Daily Camera,* for reading and commenting on my first lame effort.

Thanks to Robert H. Dott, Jr., PhD, retired professor at the University of Wisconson and Antarctic explorer, who recently examined some photos for identification. Bob and I began our careers in geology together in the late 1940s at the University of Michigan.

Thanks to daughter-in-law Sharon Sweet, RN for proof reading the manuscript.

My daughter, Anne, has many years of experience in magazine publishing and has been a great help as a consultant and contributor to my effort. We self-published a limited edition in the off chance our work was never formally published. She designed the cover of that limited edition, which has drawn raves from all who have seen it.

My wife, Mo, of sixty years and counting has been a stabilizing force and a loving smoother of rough spots and encouragement during my tenure with this effort. Oh, how she and our children have enriched my life.

In spite of all this help, the buck stops with me. I am responsible, so if there are errors or omissions, the fault is on my doorstep.

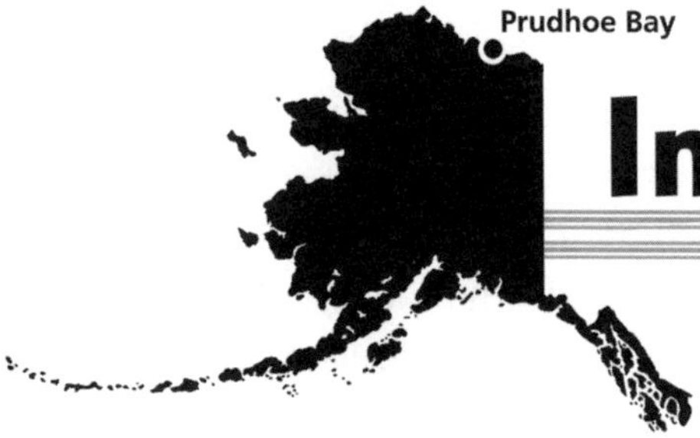

Introduction

The inside story leading to the discovery of the Prudhoe Bay oil field has never been told. When I first decided to write its history, I bounced the idea off a number of co-workers from the old days. To a man, they were enthusiastic. It quickly became evident that I was documenting a fantastic adventure during which the Atlantic Richfield Company staff in Anchorage, Alaska had discovered the largest oil field on the North American continent. At first I did not envision this as an adventure story, but once I began the research it became apparent it was.

Several scientific conferences have documented the geological, geophysical and geochemical aspects of the discovery, but no one has written the story of how the Prudhoe exploration project unfolded in the minds of the Atlantic Richfield Company (ARCO) staff. The geological and geophysical work formed the foundation, but acquiring leases, finding a drill rig and getting a budget were equally essential elements. I know the kind of dedication it took, because I participated in the huge effort invested to make the Prudhoe Bay prospect into a worthy oil exploration target.

The scientific foundation of the Prudhoe discovery was sound, but subsurface geology can be dramatically different than the geology at the surface, in unexpected and devastating ways. Many an oil exploration idea has bitten the dust, literally, because the geology of the subsurface was not accurately perceived. This was not the situation at Prudhoe Bay. The elements we thought we knew about — form, size and depth — turned out virtually as expected. In fact, several companies had similar information.

The ability of the rocks to hold oil while simultaneously allowing it to flow to a drill hole, and the degree to which the oil structure is filled, are the cruxes of all successful oil fields. They were superlative at Prudhoe Bay, and we experienced very good fortune. Some sneered that it was blind luck. Branch Rickey of the Brooklyn Dodgers said about baseball, "Luck is the residue of design," and I believe it was true at Prudhoe Bay.

While several people played leading roles in the discovery at the bay, it was a team effort in the best sense of the word. Some people were directly involved, and many more laid the geological foundation.

I am a petroleum geologist with an MS from the University of Michigan (1950). I had a thirty-five year career with Atlantic Richfield Company, one-third of which was spent in Alaska. I began my career with a predecessor company, The Atlantic Refining Company, in Midland, Texas, and gradually worked my way north through New Mexico, Wyoming, Montana, Alberta and British Columbia, arriving in Anchorage, Alaska, February 8, 1962.

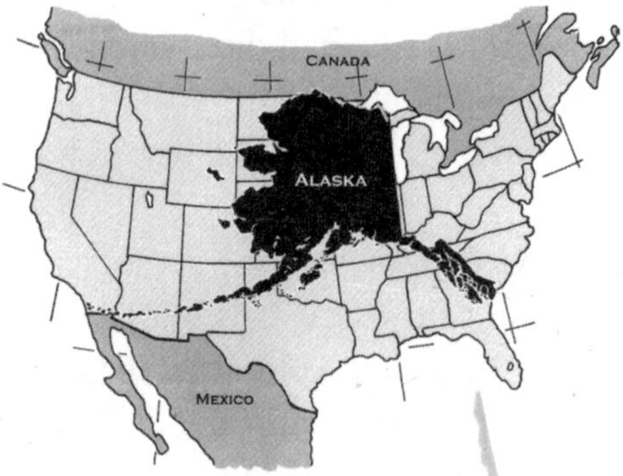

Figure 1: Alaska and the Lower 48.

The day my wife, two daughters, three sons and I set foot on Alaskan soil, my life changed forever. Prudhoe, in time, became more than a word in my vocabulary. It became the focus of the ARCO staff in Anchorage. I was District Explorationist at the time of the discovery. Very few geologists ever get to be a part of a major oil discovery, much less a world-class oil field, but we did at ARCO, and that's what I want to share with you.

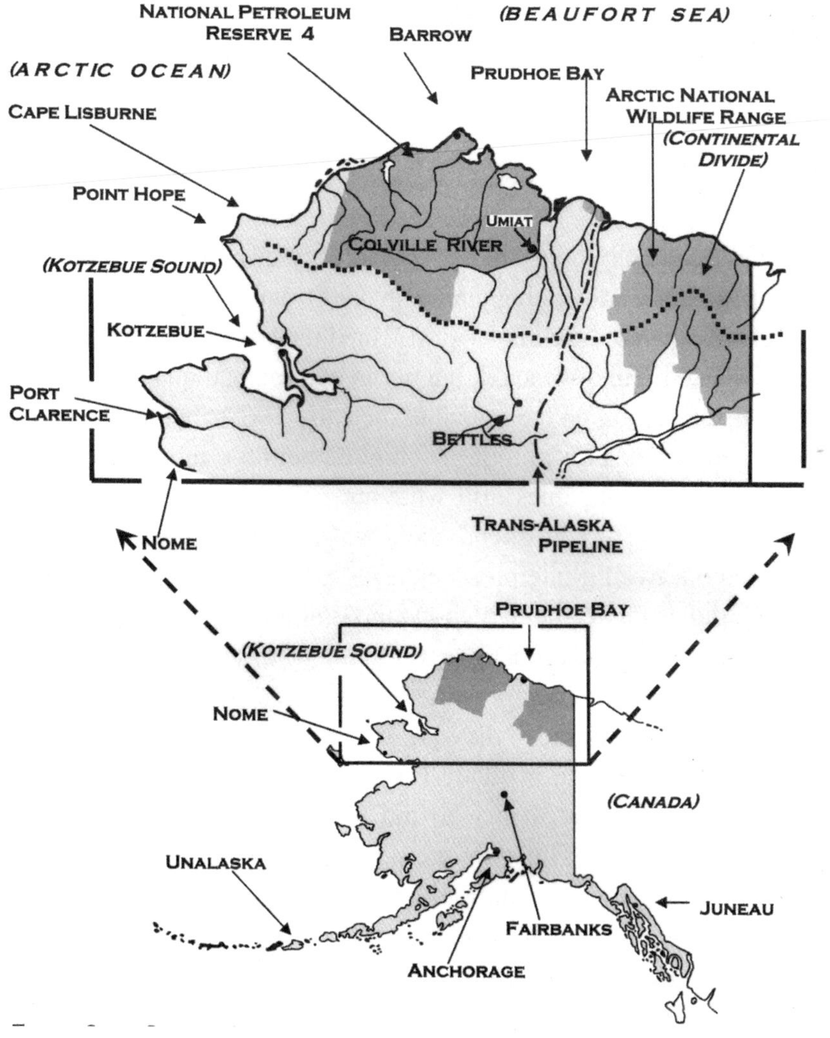

Figure 2: General index map of Northern Alaska, the North Slope.

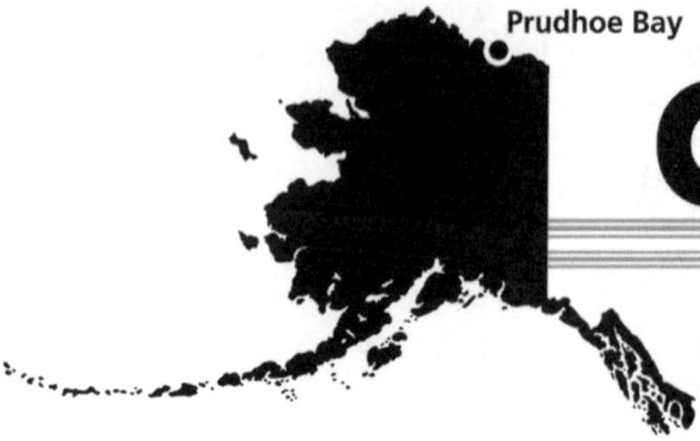

Chapter 1

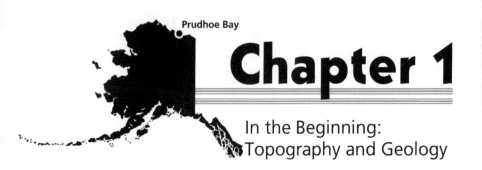

In the Beginning: Topography and Geology

Geologists and explorers studied the geology of northern Alaska over a span of six decades, beginning in 1900, with oil as the ultimate objective. Toward the end of World War II the U.S. Navy drilled many test wells and found some oil, but not in commercial quantities. Many oil companies took up the search in the early 1960s, but by the end of 1966 had spent $25 million — big money for those days. It is axiomatic that companies drill what are perceived as being the best prospects first. So, why spend additional millions on what might be considered a twelfth-rate prospect? The Atlantic Richfield Company, ARCO, did just that, and they found the Prudhoe Bay oil field, biggest in North America and probably never to be surpassed. This is that story.

It is prudent to begin with the efforts of some of the early geologic studies, because they are part of the continuum of efforts that lead to success at Prudhoe Bay. Eskimos had trod these lands for centuries, but the men who did the early work were the first scientific explorers. But why would anyone think that the Alaska Arctic Coastal Plain, commonly referred to as the North Slope, would be a good place to look for oil?

It is common for many oil-producing provinces to have oil seeping to the surface. Oil seepages occur when one, two or both natural events occur within the earth. First, oil is always associated with water to some degree, and because oil is lighter than water it is always fighting its way upward. If oil finds a route through cracks in the rocks of the earth, it emerges at the surface as seepage. Secondly, natural gas (methane mostly) associated with oil also tends to give oil buoyancy and helps carry it to the surface. Often, both water and natural gas will cause seepage.

There are two large oil seepages near Pt. Barrow at the northern tip of Alaska. Doubtlessly, it was centuries ago that the first Eskimos saw those seepages, but it wasn't until the beginning of the twentieth century that Eskimos told the trader at Pt. Barrow about them. This direct evidence of oil was big word-of-mouth news among geologists and prospectors, so the U.S. Geological Survey (USGS) probably learned about it by around 1910. These seep features were an open invitation to the first geologist who learned about them, because in his mind's eye he postulated that they leaked from an oil accumulation from within the bowels of mother earth. Like so much of Alaska's history, the exploration at Prudhoe was as a result of the search for other fortune — gold.

Gold seekers got lucky, found gold, and became gold miners. And where there were miners, there was a need for supplies, so traders opened trading posts. Traders needed the least expensive supply route. The Yukon and Koyukuk Rivers provided that route from San Francisco and Seattle to the Bergman, Alaska area. A steamship company used the rivers and established reliable transportation to Bergman, deep in the interior. This transportation link put in motion a more direct connection to the future discovery of oil at Prudhoe Bay. F.C. Schrader's USGS *Professional Paper 20* condenses this chapter from *A Reconnaissance of Northern Alaska*.

In 1896 the USGS began to think of extending geological surveying into far northern Alaska because of gold being produced in central Alaska. In 1899, Frank Charles Schrader, a geologist with the USGS, learned about the reliable Koyukuk River transportation while in Bergman. Schrader saw this good transportation avenue as facilitating geological exploration in Arctic Alaska. He proposed a program that would use the route that Lt. Howard, U.S. Navy, had followed in 1883 when he led an expedition to explore northwestern Alaska from the central interior north across the Brooks Mountain Range to Barrow, Alaska. The USGS management approved, and thus began the long involvement of the USGS in studies that culminated in the Prudhoe Bay oil discovery.

In the spring of 1900, Schrader purchased supplies for outfitting a field party. This included all the necessary camp equipment, stores and canoes because river travel was the most expedient. He bought all of this in San Francisco, a good shipping point. Everything arrived in Bergman, Alaska on time and was stored until the geological party arrived to use it in 1901.

W.J. Peters, a topographer, was the party chief, and Schrader, the geologist. There was an assistant topographer, Gaston Phillip, and five other camp hands. Probably some of them came from as far as Washington, DC. Peters and others left Seattle on February 9, 1901, arriving in Skagway, Alaska, on February 15. They had to wait three days for a snowslide to be cleared from the tracks of the White Pass and Yukon Railroad for their overnight trip to Whitehorse, Yukon Territory. Peters and his party went from Whitehorse to Bergman, a distance about 1,200 miles, using dog sleds for approximately 50 days, arriving April 10. (Bergman is "near by" Bettles, but must be nearer to Alatna, which is near the Arctic Circle, and that is the way Bergman is described in Schrader's account.)

The Peters crew loaded 385 pounds on each of two sleds. Three of the eight-man party traveled with Peters, with the others following the same route later. With these heavy loads and the number of people involved, no one rode on the sleds. Unlike today, their equipment was neither efficient nor lightweight. The freight included instruments, food for the men and dogs, and their personal gear. They slept under wolf robes.

The trip was no picnic, temperature-wise. The first day out it was –55°F, and it never got above zero for the following sixteen days. The temperature began to warm gradually in mid-March, but they still had thirteen more days with temperatures ranging from zero down to –23°F.

The first 369 miles there were roadhouses an average of one every seven to ten miles. A roadhouse was sort of the forerunner of a bed and breakfast. Meals were $1.50 each, a princely sum for then. By the end of the next 300 miles, the distance between roadhouses had stretched to twenty-five miles. The remainder of the trip saw no roadhouse accommodations at all.

In pictures, roadhouses were not prepossessing. They were log cabins, not too tall, and usually small in dimension; sixteen by thirty feet would be a large one. It is hard to imagine how they were able to take enough food for the ten men and sixteen to twenty dogs when they traveled beyond the roadhouses.

Consider the attention required by the dogs alone. The dogs' once-a-day meal had to be cooked, even if they were at a roadhouse. It was a mixture of some of the following: grain, rice, meal or flour mixed with meat, fish or grease. The working dogs were Alaska Malamutes weighing about eighty pounds. The lead dogs may have been malamutes or a mixture of any number of large dogs with good intelligence. A dog's body temperature is almost three degrees higher than a human's, and under the right conditions of size, activity and ambient temperature, dog cells may require twice as many calories per cell as humans. Calorie requirements do not increase linearly with temperature increase; it takes many more calories to sustain a dog's 101.5°F than a human's 98.6°F. Steven Ambrose, in his book on Lewis and Clark, says that some of the hardest-working men required 6,000 calories per day. Coppinger et al. report that a fifty-pound racing dog on the annual Iditarod race from Anchorage to Nome requires 10,000 calories per day. In 1901 in the wilds of Yukon, it is a wonder how they found enough food for these hardworking dogs. These dogs could have easily used 6,000 calories per day if they only worked half as hard as an Iditarod racing dog. The Peters and Schrader journal says that cooked rice and bacon provided the best source of energy for man or dog. It takes twenty cups of cooked rice and fourteen ounces of center-cut cooked bacon, or some other combination of the two, to equal 6,000 calories.

The average distance traveled was twenty-five miles per day. On one good day they traveled forty-five miles. The drivers would be riding the runners much of the time on a fast day. What about the others? Somehow they had to stay together because the sleds had the things required for survival. The journal is silent on this and other interesting unknowns. Mr. Schrader observed that the dogs did best at temperatures ten to twelve degrees below zero. That is not surprising if you learn that the larger the dog is, the more adaptable it is to colder temperatures. These days, an annual competitive dogsled race covers about two-thirds of essentially the route Peters took. The modern race is named Yukon Quest, and is from Whitehorse to Fairbanks. In that race the winner averages close to 100 miles per day, and the loser does more than sixty-five miles per day. A mail carrier on the lower Yukon River route once went as much as sixty miles. The drivers in the modern dog race, the mail carrier, as well as the men in the Peters party, all had to be in harmony with the dogs and keep them healthy.

As noted, there were no roadhouses for the final 330 miles, and at times the Peters party dug through as much as four feet of snow to reach the ground where they could stake their tents and build a fire. They used tents when possible; otherwise the dogs would want to sleep on the wolf sleeping robes. When available, they used spruce boughs as mattresses and were comfortable sleeping in the wolf robes in –30°F temperatures. They were in spruce forests during most of this portion of the route, and thus had a supply of firewood and some protection from wind.

Wet feet were one of the worst and most dangerous things on the entire trip. Surprisingly, this can happen from overflows of water on ice even at far-below-zero temperatures. The best protection, for those who had them, was the native-made *mukluk,* or boot. Overflows occur in very cold weather, because freezing of the streams becomes so intense they will freeze to the bottom of their channels. If there is still a source of water when this occurs, the water will take the path of least resistance, and that usually leads to the surface, where overflows occur. The constant supply of water makes for a slushy, mushy situation with all sorts of potential danger, for which there frequently is no warning. When the slushy evidence of an overflow occurs, dog-team drivers flaunt caution at their peril.

Over a half century later in *Two in the Far North,* Margaret Mertie described this danger when sledding in Alaska:

> Overflows are one of the most sinister of all arctic dangers. All the dog team accidents I have known about have been caused by overflow. Sometimes whole teams have been plunged into a fatal hole; many times drivers have got *(sic)* wet, and unable to start a fire, suffered frozen hands or feet before reaching help. The overflow is more deadly because it is accompanied by cold weather, and only those who have experienced it can appreciate the horror of a plunge into ice water in forty-below weather. This is another reason for the waterproof match safes and the candle stubs or can of Sterno which…mushers carry in the big front pockets of their parkas.

The Merties found themselves in such a situation:

> water [was] on the river ice, and snow with a thin slushy coating of ice on top of it. This overflow extended clear across the river, so we had no choice but to proceed. Keeping close to shore, we went splashing through a few inches of water, but as we traveled downstream it became deeper, until water lapped the canvas wrapped load, instantly changing to ice…A quarter mile further down, we came to the end of the overflow; the snow shone white and dry and safe just ahead.

Returning to the Peters party, they arrived in Bettles, Alaska, on April 14, 1901, where they rested for several days. It had cost $800 per man and $150 per ton to get the party and equipment to nearby Bergman, Alaska.

Peters studied various routes during their rest time because there were several options for crossing the Brooks Mountain Range to get to the Arctic or North Slope. Peters assigned Gaston Philip, the assistant topographer, and a Native guide to lead half of the main party in the task of scouting the Alatna River, which enters the Koyukuk about fifty straight-line miles southwest of Bettles. Philip and his party left April 23. They ascended the Alatna and an eastern branch for ninety miles. They had used all but a few days of provisions, and melt water was overflowing the ice on the streams, so Philip wisely decided it was time to retreat. First, he climbed a mountain to survey what might lie ahead and was greeted by what appeared to be a congestion of peaks in the headwater area. It did not look inviting. They could see the timberline on that stream branch, whereas the timberline was not visible for as far as they could see up the main branch. The significance of the timberline close at hand meant a steeper gradient. Nothing Philip saw was appropriate for canoe traffic.

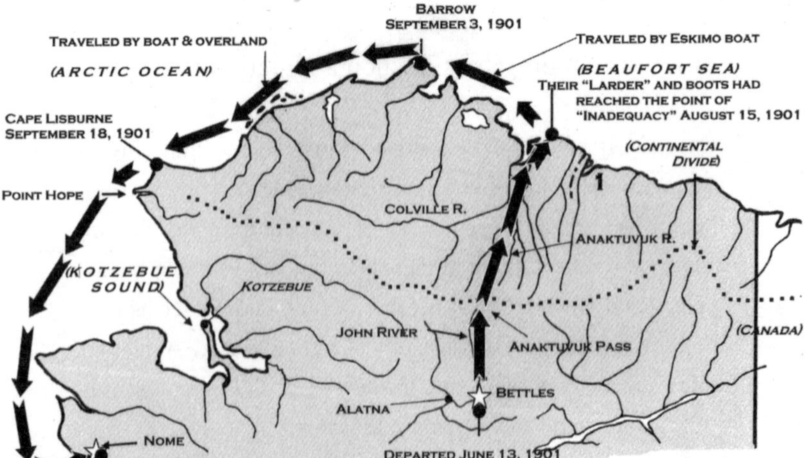

Figure 3: Peters and Schrader: the first USGS expedition, 1901.

They began the retreat on May 5, and returned to Bergman on May 10, 1901. (A USGS party in the 1920s, which will be discussed later, chose the Alatna River as their route.)

Peters also left on April 23 to explore the John River, which enters the Koyukuk near Bettles (Figure 3). They encountered all sorts of conditions: overflows, deep snow and frozen streams with bare ice. The deep snow caused them to have to tramp it down with snowshoes for several passes over the trail before it would support the sled. It was tedious but, in spite of that, they made sixteen miles that day. They met several different bands of Natives, the first one not a group, but a single woman whose three previous camps they had seen. She subsisted on rabbits she was catching with a primitive trap. On April 30, Peters had also reached timberline, but was satisfied he could see the pass that would let them get onto some river flowing to the north. They, too, retreated at that point, arriving back in Bettles the morning of May 3.

Schrader arrived during Peters' absence on the scouting trip. Peters' notes mentioned that Schrader found the trail Peters made that had been snowed in. This is the only indication that the entire eight-man party did not travel together. Schrader must have had the remaining three party members with him. The roadhouses, being modest in size, may have been the reason they traveled separately; eight at a time would probably have put a strain on the accommodations.

What was this party charged to do? They were to make a single-line traverse. That is, they would take a fairly direct route from where they were to the Arctic Ocean. There would only be time for very limited side excursions. The topographers, Peters and Philip, would use their main surveying instrument, an alidade with a plane table, to map the courses of the streams they traveled. They would use high promontories where they could see long distances, usually about ten miles apart. They would erect a rock cairn six or seven feet high at each promontory, and then they would survey lines between these various high places.

The first two cairns might be identified as one and two. When a third high place, number three, is surveyed from place number one, and again from place number two, the sighting lines cross at the location of number three. The topographer can then determine the geographic location of point three from the known geographic locations of the first two points. They would also use celestial navigation to determine latitude and longitude of some surveyed points. Where they had time, and the terrain was amenable, they would be able to sketch topographic contours. The more times a given spot can be surveyed, the more certain its location becomes.

The geologist would look for rock outcrops. Centuries of erosion caused by rain, melting snow and wind have carried surface dirt to drainage courses, large and small. This exposes rocks to varying degrees. Sometimes there are only small areas of rock exposed, other times whole mountainsides will be one giant rock outcrop. Ideally, the geologist will describe the rocks from bottom to top. Ordinarily this is from oldest (lower) to youngest (higher). He may find sedimentary rocks that are those laid down mechanically by moving water, such as sandstone, shale or conglomerates, or limestone, which is chemically precipitated from water. The other major rock type he might find is crystalline, which is formed from molten material from within the earth; granite is an example. There are many gradations in all of the rocks. If sediments are undisturbed since they were deposited, the rocks will be more or less flat lying or horizontal. Most rocks are tilted to some degree because of the way the earth has evolved. Sometimes the rocks have been tilted violently and may stand vertically or even upside down. The main measurements the geologist takes are the attitude, or strike and dip, of the rocks. The strike of the rock is the compass direction of a single horizontal (level) line on a tilted rock surface (plane). The dip of the rock plane is perpendicular to the strike and is measured in the degrees it slopes from the horizontal strike direction.

The field geologist carries a small but fairly powerful

magnifying glass called a hand lens. Depending on the situation, he might describe any or all of the following; the kind of rock (sandstone, shale, limestone, granite, etc.), the size of grains, color, thickness, fossil content, lateral continuity, the way the rock fractures or breaks, odor, the sequence of this rock type with different rock types above and below it, or anything that will help him recognize this rock if he sees it again. Another very important, but also very burdensome task, was the collection of small samples for more detailed description later in the laboratory.

To recap, Peters and Philip would plot the topography, and Schrader would describe the geology and collect samples. The five camp hands would do most of the logistics, but everyone was committed to do not only his primary function, but to pitch in with the other chores: the camp hands with scientific assistance, and the scientists helping with the logistics and camp life. The camp hands were usually very astute and practical-minded men. Many would become interested in the objectives of the parties they served and were trained to be productive helpers. Some would become knowledgeable in the fundamentals of geology and topographic work, which would increase the productivity of the party.

The scouting trips were finished in early May 1901, and then began the waiting game for break-up. This is the term given to the time and the event of the river-ice breaking and beginning its move toward the sea. In the spring, as melt water enters the channels it eventually finds its way to the bottom, and the water will become sufficient to float the ice. It is a matter of fluid dynamics, so the stream gradient and flowing water floats the ice down stream.

Peters used the time waiting for break-up to do topographic surveying in the Bergman area. Break-up finally occurred on May 29, and by June 6 the river was free of ice. The river steamer *Luella* came from Bettles and took the party and their outfits back to Bettles, where the journey north would begin.

The party left Bettles on June 13, 1901. The plan was to go from there to the Arctic coast and return by a different route. Their northerly route took them about 125 miles up the John River. This involved very rigorous and dangerous work as the large Peterborough canoes navigated through rapids and gorges. It required constant wading in snowmelt water, never above fifty degrees; however, there were no ill effects on the men. They observed that rice and bacon provided the best energy for this great physical effort.

The work ascending the John River was very hard and the progress was only three to seven miles a day with all hands working at it. Much of the time the progress came from "milking the brush" — a man in the bow of the canoe would stretch to the branches, snags and roots on shore and pull the canoe forward; another man in the stern would carefully pole forward. The water level and flow decreased, and by June 27 the river travel became easier. There was almost no time for any surveying work away from the river where the surveyors would have had to climb mountains for greater vision and, thus, more surveying coverage.

The Peters–Schrader party came to the head of the John River on July 17, 1901, after thirty-six days of travel covering more than 100 miles. Peters had arranged for a man with a horse to meet them here. The horse greatly facilitated the five-mile portage from there to a small lake in Anaktuvuk Pass, from which they entered the Anaktuvuk River on July 23.

This portage was across tundra. The geologic dictionary describes tundra as a level, treeless plain, in arctic regions. The treeless part is right, if you disregard stunted willows and cottonwoods in stream bottoms. Level it isn't. Tundra is mostly found in arctic regions, but can also be found at high altitudes anywhere. While uncommon, tundra is present in the South 48 (an Alaskan's shorthand for the continental United States). You will have seen many square miles of tundra if your travels took you over Trail Ridge Road in Rocky Mountain National Park in Colorado, or into Yellowstone National Park via the Cook City Highway from Red Lodge, Montana.

Tundra is nasty to walk on for a number of reasons. It is a mossy surface mixed with flowers, grasses and sometimes dwarf willows. The whole mass grows in an uninhibited, unstructured manner, making the surface very uneven for walking. When this is superimposed on underlying uneven terrain, it can make for difficult foot travel. Some of the vegetation will form small tufts, which grow into bigger tufts. When tundra does occur on flat ground or in low spots in the summer, it frequently will be very wet or have standing water. Shoepacks — boots with rubber feet and leather uppers — have been the summer boot of choice for years.

Tundra is a good insulator and is underlain by frozen earth. A deep bog will form where tundra is scraped away and the area is exposed to summer sun. This lesson was not learned quickly enough, and as a consequence there are some scars on the Arctic Coastal Plain. The way to prevent this is to use vehicles with jumbo tires that have very little ground-bearing pressure. Where permanent roads are needed, thick gravel pads are laid over undisturbed tundra. These roads prevent thawing of the tundra.

Anaktuvuk Pass is a major pass in the Brooks Mountain Range. The highest temperature of the entire trip occurred at Anaktuvuk Pass. It was 84°F at noon on July 24, the first of many ironies to follow. The irony is that being high in mountains usually is associated with cooler temperatures. The elevation of Anaktuvuk Pass it not all that great; nevertheless, it was the highest point on their trip and generally would not be expected to be the warmest.

The terrain from the pass changed gradually from mountainous to foothills to the almost topographically featureless Arctic Coastal Plain. The easier and faster downstream travel would have allowed for more time to do more topographic and geological work, had they not already consumed so much of the summer.

The gradient of the Anaktuvuk River and the underlying rock structures were the right combination for it to have a relatively straight course from the lake where it started to the Colville River.

The distance on Anaktuvuk River was about 125 miles as the crow flies, and downstream travel being faster, it only took them fifteen days. They stopped doing the detailed surveying previously mentioned after going only a short distance down the Anaktuvuk. This might have been because there were not the obvious promontories they had been using. It could also have been because they were running out of time. They also continued to do less detailed surveying of the river by compass. This method was faster, but less accurate. They should have been able to travel thirty to fifty miles per day based on the daily mileage Lt. Howard made on the Ikpikpuk River. If they traveled that fast they had the equivalent of ten or more days to do their scientific work. They arrived at the Colville River on August 9, 1901. The temperatures for this segment of the trip continued in the 50–60°F range.

The Peters party floated the Colville to the head of the delta in four days, arriving on August 13, after a distance of about sixty-five miles. The temperatures began to drop on that same day into the forties, then into the 30ºF range. Summer was over. They took another five days to travel another twenty miles to get to the ocean. The relatively long time for so short a distance may have been an indication they were doing more survey work.

By the time they reached the ocean, their larder had reached a state of inadequacy, and their boots were completely worn out. They met some Eskimos who were friendly enough, but generally uncommunicative. Another group of Eskimos overtook them coming from the east in a dozen umiaks, the Natives' skin boats. Peters didn't like the rough seas, and this time of year they would only worsen. He quickly concluded that their canoes would not be safe to travel the 150 miles to Pt. Barrow, so on August 20, 1901, he prevailed upon the Eskimos to take his party there. The Eskimos also supplied them with fresh fish and mukluk boots. It was a wise decision to have abandoned the canoes, because even the seaworthy Eskimo umiaks had to lay up on shore several days to wait out storms. They reached Pt. Barrow on September 3 where Thomas Brower, the whaling agent, greeted them warmly, but no mention was made of the oil seepages.

The Peters party original plan called for them to return to Bergman, Alaska by a different route, but this was obviously out of the question because they had nothing left. Their food was gone, their clothes were in tatters, they had no sleds or dogs, and winter was bearing down on them. The alternate plan called for catching a whaling ship going back south, but the last one had left a day or so before. Brower provided them with a whaleboat to sail down the coast in hopes of overtaking a vessel that would get them back to Seattle or San Francisco. They left Pt. Barrow on September 5, sailing, rowing and towing the boat. They had to seek refuge from storms for five days. On September 18, 1901, they saw a funnel and masts of a steamer in the far distance. Fortunately, the weather was good and the seas dead calm, unusual any time, much less this time of year. There were surface coal mines on shore, and this ship was loading coal for Nome, where it deposited the coal and the Peters party on September 26, 1901. From Nome they boarded a scheduled steamer for Seattle where the party disbanded.

This trek covered 513 miles in sixty-five days. That is an average of almost eight miles per day just to travel, and they were also doing geology and taking scientific measurements. There were no radios, cell phones, freeze-dried food, snow machines, outboard motors or the latest lightweight hiking and packing gear. It was just the party, their meager equipment, the elements, and their knowledge of how to travel, work and survive.

Although only a single-line traverse, the maps accompanying the report prepared by Schrader and Peters contained an unbelievable amount of information. Little could they imagine their effort was an early step in the discovery of oil at Prudhoe Bay, or that the Lisburne rock formation, named by Frank Charles Schrader, would become the primary oil objective of the *Prudhoe Bay State* No.1 discovery sixty-seven years later.

1

2

3

4

5

6

7

Chapter One Photo Captions

1. The John River in which the Peters and Schrader party waded and pulled canoes upstream toward the Continental Divide, July 15, 1901. Schrader USGS891.

2. The John River traveled by Peters and Schrader to the Continental Divide from where they traveled downstream on the Anaktuvuk River. Schrader USGS900.

3. The Anaktuvuk River in the middle distance in the vicinity of Anaktuvuk Pass. Schrader USGS958.

4. The first *Esquimaux* seen, Goobic River. (The French spelling occurred frequently in print prior to 1900.) Schrader USGS1005.

5. *Esquimaux* grave on the Arctic Coast, September 15, 1901. Schrader USGS1034.

6. Lunch stop on the Goobic River on the way downstream, August 12, 1901. Schrader USGS1016.

7. Cape Halkett Camp, Alaska Coast NW of Prudhoe Bay August 1, 1901. Schrader USGS1029.

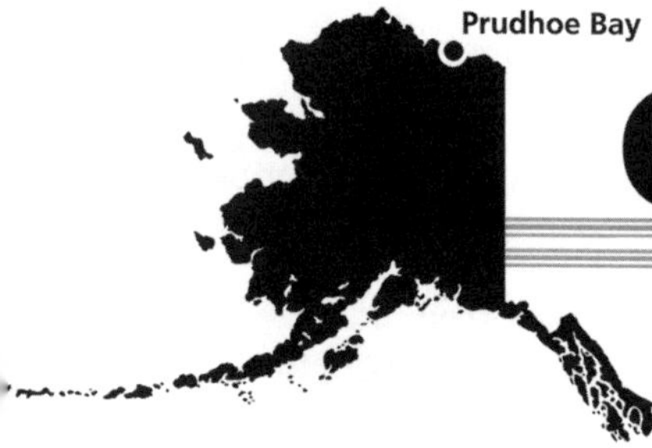

Chapter 2

Ernest deKoven Leffingwell: Hermit Explorer

Ernest deKoven Leffingwell was the next important link leading to the Prudhoe Bay discovery. He documented his work in the USGS *Professional Paper 109: The Canning River Region, Northern Alaska.* The Peters–Schrader trip was dramatic, in terms of the geological effort; Leffingwell's effort was even more so, because he did virtually all his scientific work alone. He had the help, part of the time, of a seaman and sometimes a juvenile Eskimo for logistics. Conducting geological surveys from 1906 through 1913, he worked in the eastern Brooks Mountain Range (Figure 4) and described the rock formations, layer upon layer (strata). In geological terms, this is the stratigraphic sequence on which subsequent field parties depended, and then refined, for several decades. Another unwitting player in the drama to come, little did Leffingwell know that he had named what was to become the major producing rock formation in the discovery well — the Sadlerochit sandstone.

(Author's note: Following is a description of the word "formation," a term that will be used frequently throughout this narrative. Formation is the designation geologists give to the primary rock units they find worldwide. There are no minimum or maximum thickness requirements for the size of the area of exposure. The rocks must be distinctive enough to be represented on a map. For example, if a geologist finds rocks that are 100 feet thick, sandstone, and red, with limestone above and below the sandstone, and he can see it in several different places, that would be enough information for the formation term to be used. Formations can be any geologic age and may cross age boundaries, and thus one part may be one age and the other part another age. When several formations are lumped together, this is called a group. If there are variations within a formation, each distinctive variation will be called a member.)

Dr. Leffingwell was an icon to all who came after him in the study of the geology of Arctic Alaska. Unfortunately, the general public never recognized his fame in geological circles and his tremendous contributions to that science and the geography of Arctic Alaska. He worked with explorers Ejnar Mikkelsen of Denmark and Vilhjalmur Stefansson from Canada, who also explored in northern Canada and Alaska. Leffingwell's papers and journals are with the Stefansson collection in the Dartmouth College Library. Leffingwell's work was more localized and in greater detail that that of explorers Peary and Amundsen and, thus, was overshadowed by their expansiveness.

Ernest Leffingwell was the descendent of a long line of Episcopal clergymen, and it was family expectation he would follow in that calling. He had a classical education beginning in his father's St. Mary's School in Knoxville, Illinois, and at age ten was sent to Dr. deKoven's School for Boys in Racine, Wisconsin. From there, at age sixteen he went to Trinity College in Hartford, Connecticut to prepare for a career in the Episcopal Church. He graduated with a BA degree in 1895 and an MA in 1898. He then attended the University of Chicago to do graduate work in the years 1897–1900 and 1903–05. Nothing in the record indicates when he became interested in science, or the degree objective at Chicago, but it is

Figure 4: Leffingwell's self-funded expedition 1906–12.

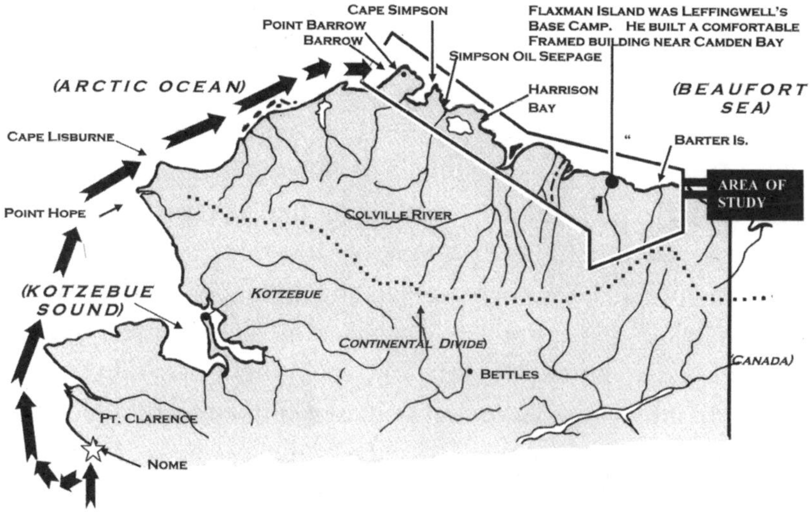

likely that he did all the course work for a doctorate in geology and science. He worked on writing a thesis but never completed the requirements for a degree at Chicago. It became obvious from his future work that he was very knowledgeable in the fields of geology, astronomy, surveying and celestial positioning. Trinity bestowed on him an honorary Doctor of Science degree in 1923.

He had always been interested in arctic exploration, but two events coincided to make the interest a reality. In 1900 there was a lecture at the University of Chicago by a respected arctic explorer. This raised his passion for exploration even higher, but the real key came shortly after the lecture when he heard about the formation of the Baldwin–Ziegler Polar Expedition. Leffingwell applied for membership, and not only was he accepted, but he was made director of the scientific staff. He attributed this to the good recommendations he had from the geology faculty at the University of Chicago. He was directed to draw up plans for the most comprehensive scientific work that had ever been attempted in the polar region. This is as close as Leffingwell ever came to saying what the objective of the expedition was. Ejnar Mikkelsen became Leffingwell's chief assistant, a lifelong friend and later an exploration leader in his own right.

The expedition departed from Dundee, Scotland on one ship in 1901. From there they went to Tromso in extreme northern Norway where two other ships joined the first ship. At that point the flotilla went to Archangel, Russia, which is on the Barents Sea, a part of the Arctic Ocean. There they took on 400 Siberian sled dogs, six Siberian ponies and several Russian dog drivers. The next stop was 1,000 miles farther north to Alger Island on the south end of Franz Joseph Land archipelago where they spent the winter. At this point they were only eight degrees of latitude from the North Pole. After transporting tons of supplies 100 miles farther north, overland, in the spring of 1902, they returned to their ship and thence to civilization. Years later Leffingwell made the comment several times that he and Mikkelsen should write about this aborted effort, but in the same breath indicated they had no explanation for the sudden abandonment of the effort except to say that Ziegler replaced Baldwin as leader.

The experience brought Mikkelsen and Leffingwell briefly together. As Leffingwell later described it, they "hit it off." During their time together on the Baldwin–Ziegler expedition they discussed the possibility of exploring the Beaufort Sea north of Alaska for land reported by whaling captains and Alaska Eskimos. Leffingwell returned to the University of Chicago until the fall of 1905.

Even before the brief aborted trip with Baldwin and Ziegler, Mikkelsen had sought berths on Solomon Andree's balloon expedition to the North Pole, and Barron Eduard von Toll's fossil hunting expedition to the New Siberian Islands. He missed these "opportunities" because he was only sixteen. Andree's balloon, the *Eagle,* went down and his party's remains were not found for thirty-three years; and von Toll's remains were never found. Mikkelsen grew up wanting to be an explorer.

In 1905 Mikkelsen began to solidify plans for his and Leffingwell's desire to explore the persistent rumors of land — and marine current information that indicated its possibility — several hundred miles off the north coast of Alaska. Mikkelsen was a twenty-four-year-old sailor, and not a scientist, in 1905. But, so what? One didn't need to be a scientist to prove an interesting piece of land existed, which, as he later wrote, he had made up his mind he was going to discover.

He began knocking on doors in Denmark with the goal of raising 50,000 crowns, thinking he would find fifty people with 1000 crowns each. He found one. He had a personal connection to the Royal Geographical Society in London, so he went there and had an audience with Queen Alexandria, who didn't promise money but agreed to be a patron. Mikkelsen doesn't say in his book, *Mirage in the Arctic* (written fifty years after the trip) how much he finally netted in London; but it was enough, with what Leffingwell's father matched, to lay the groundwork for exploration, so he returned to Denmark to buy equipment for the trip.

In London, the Royal Geographical Society had recommended that Mikkelsen see the American Geographical Society, so his quest for additional financing took him to the United States. He met with President Theodore Roosevelt, who asked that Mikkelsen take possession of any newly discovered land for the United States. This upset Russia, but that proved to be a door opener because of the attendant publicity. He was subsequently put in touch with many of the movers, shakers, and organizations of the times, such as John D. Rockefeller, the railroad magnate Harriman, Alexander Graham Bell, and The Carnegie Foundation, among others. President Roosevelt gave permission for U.S. revenue cutters, in the areas where they sailed, to support Mikkelsen's exploration. The President also "greased the skids" with the customs people to process Mikkelsen's equipment from Denmark and Norway through U.S. Customs with no red tape. By the time he left New York, Mikkelsen had an additional $8,000. On the way west he met with Leffingwell to renew their friendship and make additional plans. They were happy to learn that Leffingwell's father added $10,000 to their budget.

In Vancouver, BC, Mikkelsen found a 66-ton schooner originally well constructed with conventional expensive tropical hardwood. She had a checkered history, beginning as a Japanese naval vessel, then was used for marine mammal poaching in the Pribilof Islands, pearl fishing in the tropics, and opium smuggling, when her owners ran her aground with the law close at hand. Mikkelsen purchased her from a Canadian who had repaired her and put her to legal sealing.

As one might expect, it took much more than expected to make her ready for the adventure. Mikkelsen wasn't an experienced money manager and was soon notified by their banker that he was overdrawn…to the tune of $12,000! They went back to the proverbial "well," with Rockefeller wiring $5,000 and Leffingwell's father coming through for "more than he had before." They wanted to divorce the little ship from her nefarious past, so they decided to change her name from *Beatrice* to one of their stalwart patrons. Queen Alexandria had put Mikkelsen in touch with Mary, Duchess of Bedford, so they named their ship in the Duchess's honor. Canadian law prevented Leffingwell or Mikkelsen from being the owners of record, so they enlisted the Duchess to take ownership, which she did, and then leased it back to them for six shillings per year. There was a christening and departure party on board, which was attended by many of the "top ten" of Victoria. They sailed from Victoria in the spring of 1906.

They made themselves joint commanders, but Mikkelsen had done most of the organizing, and was deemed the captain of their ship. Their meager funding precluded having an engine on the ship, so they were severely limited in their ability to navigate to best advantage in the ice. Their cruise north was routine (after they discovered a serious leak, which should have been fixed, but they were broke again. There was a hint of the past in the form of a bilge pump, so it was employed at regular intervals. A particular tack was worse, so they avoided that as much as possible.) Routine that is, until they got near the gold fields of Alaska when it became obvious the crew had plans to jump ship. Nome was the port of choice for the would-be gold miners, and the reason for their relaxed attitude about the wages before leaving Vancouver.

At Port Clarence they met the U.S. revenue cutter *Thetis,* whose skipper told them to keep an eye on the crew or they would desert at the slightest opportunity. Mikkelsen thought there was little danger, as Nome was seventy miles by water or over the tundra, but, sure enough, the next morning they discovered the crew had taken the dingy. An appeal to the *Thetis* brought help, and the *Thetis's* captain sent his crew to bring the deserters back to the *Duchess*. The captain of the *Thetis* told Mikkelsen to go to Pt. Hope and wait there for them.

When the *Thetis* arrived at Pt. Hope, the captain said he had four men who would be glad to sail aboard the *Duchess*. In the meantime, Storker Storkersen, an excellent seaman, had seen the error of his ways and decided to stay with the ship. The captain of the *Thetis* held court, and the other three were tried and convicted of desertion. They were sentenced and put ashore in an Eskimo village of about 400 individuals and left to fend for themselves, hundreds of miles from the gold fields.

From Pt. Hope it was a slow journey of wait and go because the ice was so tightly packed to the shore on to the north and east. They were still at Pt. Barrow at the end of August. The captain of the *Thetis* had arrested a whaling captain on the charge of misbehaving with Eskimo women onboard ship. He agreed to be lenient with whaling captain if he would tow the *Duchess* to the east, and that is how the Anglo–American Polar Expedition finally got to what became their winter quarters at Flaxman Island.

After spending a comfortable winter eating lots of fresh salmon with the Eskimos, in the spring of 1907 it was time into go to work to look for the rumored land north of Alaska. Mikkelsen, Leffingwell and Storker Storkersen started their trip around mid March 1907, over the Beaufort Sea Ice (Figure 4.1). Some Eskimos accompanied them to help negotiate the very rough ice that had piled up near shore and the open leads also close to shore. In a few days the rapid drift of the ice to the west versus the predictions of static or nearly static movement, which favored the existence of a yet-to-be discovered land mass, forced them to abandon the ideas on which they had placed so many hopes.

One wonders how they survived the rigors of the trip on the ice. They had to be miserable most of the time. The first thing they encountered were illusions of mountains in the distance. It occurred repeatedly. Huge blocks of ice would rise under them time and again. Then the ice blocks might submerge, the fear being that the men would be carried with them. Open leads of water had to be crossed. For a while they would get aboard small flows and hope the currents would go the right way. Occasionally they would go the desired direction, sometimes they would stall for long periods, sometimes the flows would capsize. After seeing some of these, they gave up on this method and changed to wrapping a sled in canvas to use as a ferry between flows. Had they ever lost one, they would have been in dire straits from the loss of vital gear. Rabies had infected most of the dogs, and the devastating effects came and went. They had inoculations of some sort, but eventually just ignored it. At night their sleeping bags were frozen with condensation and they would have to force themselves in little by little as their body heat thawed the bags. The first day of a given storm was a welcome rest, the second day was too much, and the third hard to bear. More than once, ice split under the tent and there was a panic until all was safe. Once it cracked between two sleeping bags. Finally after two months they had reached the critical point of not wanting to drift west of Pt. Barrow. Had that happened they would have been adrift until they perished. It was learned much later that west of Pt. Barrow longitude the currents

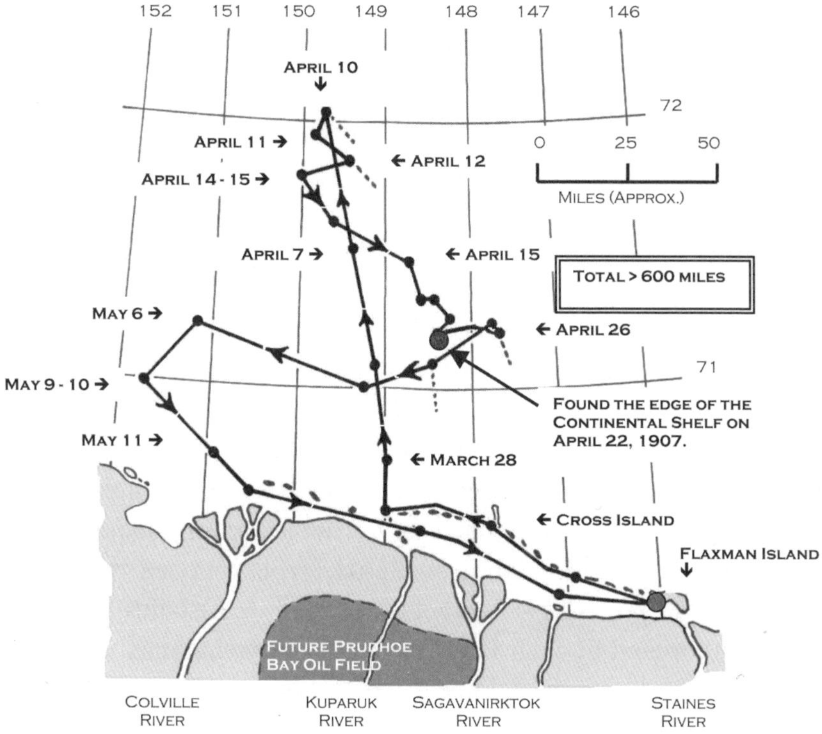

Figure 4.1: Leffingwell Expidition
Mikkelsen and Leffingwell left their Flaxman Island base March 15, 1907 and returned May 15, 1907. This is a location plot of the Mikkelsen and Leffingwell two-month trip over the Arctic ice pack while looking for the elusive rumored land North of Alaska. Mikkelsen's account forty years after the fact indicates they traveled farther north and drifted farther west than is documented on this map (which the author believes takes precedent over Mikkelsen's memory). They were about 100 miles north of Alaska and 150 miles northwest of their Flaxman Island base. They appeared to have traveled over 600 miles.

turned north toward the North Pole. By this time the remaining dogs were so wasted that they no longer even had strength to snarl at one another. The equipment was all badly worn, and much had been discarded. After seventy-odd days they struck out for land. They were over 200 miles from Flaxman, but in ten days they were back on land just twenty miles east of Pt. Barrow. In another ten days they were back to Flaxman Island, approximately May 19, 1907.

They had taken numerous soundings of the ocean floor and established the location of the continental shelf, significant information but of no great value beyond its being an addition to accumulated geographical knowledge.

Was it foolhardy? No, but disappointing to Mikkelsen and Leffingwell. They had only seen sea ice, but no islands in the area of the Beaufort Sea, which they had traversed. Much later, people flying airplanes navigated the rest of this vast wilderness. In August 1946 Col. Joseph Fletcher of the USAF discovered a huge ice island of 350 square miles. A rectangle thirty miles by twelve miles is 360 square miles. A picture revealed it to be oblong and not rectangular so it may have been somewhat longer or wider than the thirty by twelve dimension to make up for its rounded corners. It was 100–120 feet thick. Although not land, ice islands could be mistaken for land by their land-like topography, and may very well have been the sightings claimed by early whaling ships and natives, which were the source of the rumors of land that Mikkelsen and Leffingwell were chasing. Fletchers "Island" was spotted 250 miles NNE of Pt. Barrow. Others have been found since then. They drift slowly in a clockwise direction, and when they get northwest of Barrow they drift toward the North Pole. In light of this later knowledge, it was logical that people with inquiring minds thought it was worth looking for land, and those individuals were Mikkelsen and Leffingwell.

When Leffingwell and Mikkelsen returned from searching for land, they found that the movement of the sea ice had damaged their ship beyond repair, so they salvaged their supplies and much of the wood from the ship, with which they built a cabin on Flaxman Island. Leffingwell had everything he needed so he decided to stay. Mikkelsen said he would too, but Leffingwell encouraged him to pursue other plans he had for exploring

Greenland. Mikkelsen became an authority on Greenland and was eventually appointed Inspector in East Greenland, 1933–1950. He was a recognized authority on the Polar region in general, and did much writing and speaking on the subject.

When Mikkelsen decided there was nothing more he could do in Arctic Alaska, all he had to his name were the remains from the Beaufort Sea exploration on the pack ice that he and Leffingwell had done. He debated and studied the shorter overland route to Fairbanks, but only a few scattered Eskimos inhabited it. He calculated he could not carry enough to support such a long trip alone. He then decided on the coastal route from there to Nome. He met another Norwegian from a whaling ship that was wintering east of Flaxman Island, and traveled with him to the village of Barrow. The trader at Barrow gave him supplies to continue from Barrow to Nome. Leffingwell and Mikkelsen had a good relationship with the friendly and helpful local natives near Flaxman Island, but when he got to Wainwright, and supplies were running low, he found the Eskimos there had exchanged many of their good Eskimo qualities for white man's selfishness and other undesirable habits. Other places, the Eskimos were in as poor straits as he was. He was destitute, but continued grubbing his way south. When he was closer to Nome he began to encounter gold seekers, many of them unprepared for the climate. He tried to encourage three men to turn around and travel back to Nome. They wouldn't hear of it. He traveled a little farther and joined another group going to Nome. They decided to backtrack and persuade the men he had passed to relent and return with them, but only found their frozen remains.

When he eventually reached Nome he was treated like royalty because they had heard, and believed, that he and Leffingwell had perished on the Arctic ice pack. He was the guest of the city, with much recognition and many parties. Just as Mikkelsen was leaving Nome, a merchant confided that he could not get more goods unless he could find someone to take his gold from Nome to Seattle.

Mikkelsen reluctantly agreed, but he was exposing himself to possible banditry, and he hardly needed a heavy load of gold.

From Nome he headed overland to Fairbanks following the U.S. Army telegraph line. The trail was obscure because it departed from the telegraph wire from time to time, and snowfalls between travelers did a good job of hiding it. Many of the roadhouses along the trail were nothing but filthy hovels. He traveled alone until, the last few days before reaching Fairbanks, he was joined by a man who, having found enough gold to buy a farm, was returning to his family in the Midwest.

In Fairbanks, a much larger and affluent city due to gold production, Mikkelsen was treated even more royally than in Nome. In spite of his well-worn skin clothes, he was accepted and became the excuse for what must have been endless partying. (He met unexpectedly with his farmer friend from the trail who, sadly, had lost all his nest egg to the local ladies of the evening and bartenders.) Being eager to continue with his trip, Mikkelsen was able to ride on the mail stage from Fairbanks to Valdez.

He had departed from Flaxman Island on October 15, 1907 and this part of his journey ended the following year about April 1; he estimated the effort covered 3,000 miles.

Mikkelsen left Valdez via steamer. The skipper, also a Norwegian, for some reason felt he had to prove himself to Mikkelsen, and proceeded to show his passenger how he could negotiate a sub-surface rock about sixteen minutes from cast off. He had Mikkelsen be the timekeeper for the maneuver. All did not go well; the ship struck the rock and stayed put. Everyone was safely evacuated and the ship went to bottom three days later.

A subsequent steamer took Mikkelsen to Seattle where on his arrival he was immediately paged. He knew he would be met by someone to collect the gold he was carrying, so, to inject a bit of drama into the situation, he initially ignored the pages. Ultimately he identified himself to the pager, a man carrying a large suitcase, and finally unburdened himself of the gold (later reported to be worth $40,000!). The person rushed away without so much as a thank you, much less any gratuity.

As an aside, nearly 100 years later a young man from Fairbanks, Ned Rozell, and his dog, Jane, walked the Alaska pipeline route from Valdez to the Arctic Ocean over a period of two summers. He documented his trek in *Walking My Dog, Jane.* Mr. Rozell had modern equipment and lots of support along the way, but even this trek was not all a walk in the park.

To return to Leffingwell, he was not a hermit in the sense of being reclusive. The early years he did live alone, but had the assistance of an Eskimo family. At a later time, a seaman who joined him on a return trip from the United States lived with him. Leffingwell was, however, a hermit as far as his work was concerned. He had little or no help with his scientific efforts.

The elapsed time from Leffingwell's first departure from Seattle until his last return from Alaska was about eight years. He was either traveling to or working in Arctic Alaska for seventy-eight months. He estimated that of those seventy-eight months, he was in the field — that is, away from the relative comforts of his Flaxman Island headquarters — for thirty months. He also estimated that he had made camp in the field over 380 times and had traveled 4,500 miles. The field trips covered an area seventy by seventy miles. Those 4,900 square miles are part of the pristine Arctic National Wildlife Range. While it was happenstance that he located at Flaxman Island because that is where ice trapped his ship, it turned out to be fortuitous because he was well located in relation to his first love of Pleistocene-age geology and outcrops of older rocks that were later found in the subsurface at Prudhoe Bay.

Neither when interviewed, nor in his writing, would he advance the idea that it was dangerous to live and work in the Arctic. He said there are lots of places someone can freeze to death. There was danger once when he and an Eskimo were boating and his dory broke loose. The only recourse was to dive into the water and retrieve it, which he did. He described it as a life-and-death situation. Did he mean losing the dory, or diving into ice water to bring it back? Death from either outcome was possible. Although he does not say so, there is a substantial likelihood that he would have died of exposure if the Eskimo had not been there to help him build a snow shelter and boil tea to restore his body temperature.

Leffingwell went to Seattle in September 1908 to buy supplies for another three years. He learned there would be no whale ships going to the Arctic Ocean in 1909, so he decided to have a fifty-foot yawl built, equipped with a twelve-horsepower gasoline engine. He was skeptical of the cookstove installation, but the contractor assured him it was customary, safe and sound. Leffingwell's acceptance of this explanation was a mistake, because the builder had cut corners to recover the increased cost of lumber after the contract was signed. Instead of renegotiating, Leffingwell and three sailors left Seattle in his new ship, the *Argo,* on May 25, 1909. The budget was tight and his wage scale attracted three men whom he discovered later were in the depths of alcoholism.

Leffingwell had done some recreational lake sailing when younger, but he was about to tackle the North Pacific Ocean and the Gulf of Alaska, which can have severe weather matching any in the world. He had not practiced navigation, but he trusted that his knowledge of astronomy and surveying would get them back to Flaxman Island. The party immediately discovered that the engine exhaust was too close to wood and was smoldering. This was easily fixed by adding some tin flashing at the appropriate place. Soon after that, they found that the stove was not sufficiently isolated from the structure of the boat, as Leffingwell had feared. It actually ignited the substantial bulkhead, which had to be immediately doused with water to forestall disaster. He didn't believe he could spare the time to return to have the problem corrected. The weather was terrible. The boat was too frail to buck the waves, so they spent many days just riding the waves trying to remain afloat. They were repeatedly getting soaked and were blue with cold, with no means to dry their clothes, warm themselves or cook.

The crew, not the best natured anytime, became quarrelsome, with the two who were then off duty getting into a knife fight. Leffingwell separated them without anyone getting hurt. This was the only occasion of complaint of a hardship, and it was an embarrassment to him later. The fight had been the last straw, and he was so cold, wet, hungry and generally miserable, he was ready to quit. In desperation he decided to try the stove again. He was more ready to burn than freeze or starve to death. He started the stove, and they had to throw water on the wall behind the stove almost continually to keep it from burning, but the crisis was over. They got warm, dry and fed.

Toward the end of June he had been on duty forty-eight hours without sleep, wondering about his ability to navigate. They were in fog. He believed they should have been near land. Gradually the fog and mist cleared and dead ahead was Unalaska, in the Aleutian Islands, just where he hoped to be. On docking, other problems faced Leffingwell. He gave the crew shore leave, and not only did they come back in a drunken state, one lapsed into the DTs *(delirium tremens)* and started breaking things and throwing furniture overboard. Leffingwell said the process of physically subduing him was the fight of his life. The captain of a nearby U.S. revenue cutter invited Leffingwell to join them for dinner. He was going to have another of his crew row him to the cutter, but this man, too, turned out to be inebriated and fell overboard. Leffingwell did what he had to do and rescued him. The sailor was dead drunk, so he did not cause any additional disturbance. By the time Leffingwell got himself and the drunk into dry clothes and the drunk put to bed, he had missed his dinner invitation. Exhausted, he fell into bed and closed his eyes for the first time in sixty hours.

The whale ship *Karluk,* whose captain Leffingwell knew, was also in Unalaska, and he offered to tow the *Argo* to Nome.

Misfortune intervened when the towline split in a storm along the way. At Nome, delayed again by poor weather, he had more trouble with the drunken sailors, but he needed them for some difficult passages. Farther up the coast at Point Clarence, he finally left two sailors behind for again ending up in a stupor. On finally arriving at Pt. Barrow, the remaining sailor, who had tried to burn some construction lumber for fuel, left as well. A crewman Leffingwell knew from the expedition with Mikkelsen replaced him, and he hired Scotty McIntyre, who stayed with Leffingwell for three years, but was not involved in any of Leffingwell's scientific work. They landed at Flaxman Island August 23, 1909, and then made a trip back to Barrow to bring supplies that had been shipped there. He remained at Flaxman Island (see Figure 4) until September 1912.

The next three years were productive ones for Leffingwell. Readers are reminded that no one asked him to do what he was about to undertake. He was not a shill for some rich tycoon who wanted information gathered. He was going to be doing scientific work of his own volition, and was underwritten by private sources, probably mostly from his father. Alfred H. Brooks, then the director of the U.S. Geological Survey, describes Leffingwell's work. "Nearly all parties that have undertaken exploration in Alaska and Polar Regions have been large enough to permit both the scientific observations and the physical labor incident to travel to be divided among several men. Not so with Mr. Leffingwell's party, for most of the time after the departure of Mikkelsen in 1907, he had only one white man to help him, and he, one who could take no part in the scientific observations. In fact he made many of his journeys with only one or two Eskimo companions and he made some entirely alone." (On many occasions, the Eskimos would depart after arriving at the site to be investigated.)

The only hint Leffingwell gave as to his possible motivation was his interest in Pleistocene geology, which is the geologic time period we are currently in and covers the last million years. In spite of the fact that the Pleistocene was his first love, he gave thorough attention to all geologic ages, and significant time to determining the latitude and longitude of important landmarks. He also made many marine soundings when he traveled along the coast. Geologically, he described the sedimentary rock outcrops in terms of thickness, composition, sequence and correlation from one place to another. The geologic term for this is stratigraphy.

Pleistocene geology in this area is dominated by permafrost, which is permanently frozen ground made up of soil, vegetation and ground ice. Ground ice, more or less clear, is a very substantial portion of permafrost all across the Arctic Coastal Plain. It can take many forms, from vertical wedges to lenses, to continual masses covering large horizontal areas, and all shapes in between. Alfred H. Brooks again on this subject, "Not the least of Mr. Leffingwell's contributions to science is his detailed study of the ground ice." This is the phenomenon on which Leffingwell probably spent most of his effort. It provided an excellent starting point for Arthur Lachenbruch who completed the dissertation for a doctorate at Harvard in the mid-1950s on the subject of ground ice. Leffingwell's studies withstood the test of time.

Through the years the United States Geological Survey has been meticulous in adhering to the highest scholarly standards. It speaks volumes for the quality of Leffingwell's work that the USGS would provide office space to Leffingwell and publish the results of his fieldwork as *Professional Paper 109: The Canning River Region, Northern Alaska.* An amazing fact is that all the expenses for nine summers and six winters were only $30,000, and half of that was spent the first year when he was a member of the aborted Anglo–American Polar Expedition. If the remaining expenses were distributed evenly over the other eight years, he was spending less than $2,000 per year.

He established a fairly comfortable headquarters camp at Flaxman Island near the mouth of the Canning River. The island is three miles long and half a mile wide with a maximum elevation twenty feet above sea level. In his report he gave many suggestions about how to make shelters, from temporary to permanent. Modern materials have obviated many of the techniques that worked for him. Temporary field quarters — that is, some form of tent — have not changed much except for the fabrics used. Leffingwell believed one of the most efficient temporary structures in his time was an Eskimo tent made of willow sticks and skins. The willow sticks were used to form an arch nine feet in diameter and five to six feet high at the center. This was covered with two layers of skins and well banked with snow around the base. Five people sleeping in a well-made structure like this would keep the inside temperature fifty degrees above the outside temperature. The best sleeping bag was made of sheepskin with the wool inside. A caribou hide beneath, with the hair down, kept the sleeping bag dry and increased its warmth. Leffingwell's feet got cold only once when the temperature was –50°F. He built his own dog sleds and experimented until he found the optimum dog harnesses.

The Peterborough canoes used by Peters and Schrader and subsequent explorers for river travel were too easily swamped in the saltwater lagoons and open sea. The Eskimo umiak was ideal because of its carrying capacity, lightweight and shallow draft. Leffingwell's most effective boat was a dory. These have a pointed bow, flat stern and flat bottom, but with a centerboard. His was light enough that he could pull it onto the beach by himself even though it was twenty-seven feet long.

Cold climate clothing also has advanced and made obsolete that worn by Leffingwell, but readers may experience vicariously the clothing he wore and was perfectly happy with. Leffingwell thoroughly documented what it took to live in the Arctic in his USGS *Professional Paper 109.*

> On the upper part of the body a light fawn skin shirt is worn with the fur next to the body. A similar shirt is worn over this one with the fur outside. The fur shirts are furnished with

> hoods with a fringe. This fringe is supposed to keep the wind from the face, but fashion is the chief reason.
>
> The shirts should be large enough to allow the arms to be pulled inside to warm the hands or make it more comfortable to wait out a storm in a temporary snow house. Fur breeches of shorthaired adult caribou are preferable, which may be either full length or reach below the knees. Fur stockings are knee length or shorter, according to the length of the breeches. Like the breeches the hair should be inside.

A complete set of furs like this would weigh eleven or twelve pounds.

The weather offered some surprises. During Leffingwell's time the coldest months were January to March, when the usual coastal temperature was –20°F. About a dozen times a year it would get as low as –40°F, and for three days one February it got to –60°F. February would occasionally have a warmer spell. One time, in February 1911, there was an unheard of, unusually warm, maximum temperature of 44°F. This prevailed for ten days, and it even rained. Leffingwell could not travel because one of the rivers he needed to cross had more than a foot of water on top of the ice. A short time later it was a more usual –50°F.

Drizzle was common, but it almost never rained hard. Most of the snow fell in April and May, but all winter there was seldom more than six inches at a time. During June to August, temperatures varied between 30°F and 50°F with a maximum of 70°F. The climate is arid. Persistent thirty-mile-per-hour winds were common, fifty miles per hour was not unusual, and once there was a gale of eighty-four miles per hour when they were on a trip away from Flaxman Island. The sun set about November 20 and would rise in late January. During this long night there was about six hours of twilight sufficient for travel.

Currently (July 2007) there is still discussion of drilling for oil in the Arctic National Wildlife Range (ANWR), and there is also much concern about the effects this would have on the caribou population of the Porcupine herd that frequents part of the Arctic Coastal Plain within the ANWR. However, according to Leffingwell's notes, the caribou herd declined rather dramatically a couple years in this pristine area, in the absence of any drilling activity. From Leffingwell's *Professional Paper No. 109,* "Formerly, they [caribou] could be seen in numerous large herds scattered over the tundra, but within a very few years, they have become much reduced in numbers. In the spring of 1907 so many haunches were offered to the writer's party they were finally refused…In 1912–14 not over half a dozen haunches were procured. At present, caribou are reported to be fairly abundant on the Yukon side of the divide." Leffingwell attributed this to the caribou being rounded up by the natives "and killed or driven back into the mountains." In all likelihood some other dynamic was in play, because Leffingwell reports that the 1910 census put the entire Eskimo population between Pt. Barrow and Canada along the coast at sixty-five, of all ages, covering a distance of more than 500 miles. In addition, Leffingwell remarked that Eskimo marksmanship success for seals was less than 10 percent. One might reasonably believe caribou are faster than seals, making marksmanship less than 10 percent for them. Eskimo hunting pressure does not seem to be the answer. One might wonder then what caused the scarcity of caribou. Getting an answer would be a guessing game: maybe parasites, maybe some disease, or maybe they just went somewhere else. One thing about caribou numbers is certain; their numbers are constantly changing, sometimes dramatically, and the wildlife people seldom have answers for these crashes and explosions of animals.

Polar bears were likewise not abundant. In his time there, Leffingwell saw only two, and they were more than twenty-five miles from shore. Eskimos in his area annually killed about a dozen sows that migrated to high banks along the shore to drop their cubs.

Game birds, however, proliferated during Leffingwell's stay. Eider ducks were seen by the tens of thousands. One year they estimated 700,000. Eskimos estimated a million. During heavy flights a single hunter could bag 400 to 500. Black brant ducks, white-fronted geese and old squaw ducks were also numerous. Ptarmigan were plentiful, but in far less abundance than the waterfowl. All of the game birds were a source of food for Natives and Leffingwell. He did not mention snow geese.

A 2½-inch gill net would catch more fish than could be eaten during the periods when fish were migrating. When the fish were not migrating, Leffingwell would never catch more than a dozen.

Leffingwell had such a ready supply of game animals, birds and fish that there was no need to bring canned varieties, except for bacon and salt pork. It was fortunate that flour was not easily damaged because it was used in great quantities. Oleomargarine was not as good as butter, but was better than most of the butter brought to Alaska. He liked dried milk, but evaporated eggs were not palatable. Leffingwell had a sweet tooth for jams and jellies, so he kept those well stocked.

In the spring of 1910 Leffingwell went whaling to help pay expenses. A whale would have netted him $10,000, but they didn't see any. He accumulated furs for several years, which netted him a few thousand dollars. He discussed his trips with the same air of nonchalance that most would use about running down to the grocery store, yet many of those trips were significant, some of them in sub-zero temperatures.

In the fall of 1912 Leffingwell went back to Washington, DC, for four months, where the USGS provided him an office and where he began to write what ultimately was finished as *Professional Paper 109.*

He went back to Flaxman Island on August 18, 1913 with many heavy cases of better instruments, counting on Eskimos to help bring them ashore in a large dory he had also brought with him. The

Eskimos historically gathered about August 1 to trade, but had already departed. Leffingwell was faced with continuing his work with the old instruments or giving up. He had all the other supplies he needed, so he stayed to do what he could. This return trip was made possible by acting as guide to the Canadian Arctic Expedition under the leadership of Vilhjalmur Stefansson and Dr. R.M. Anderson. (Leffingwell, in *Professional Paper 109,* does not go into any detail about the Canadian Expedition, but Internet information on Stefansson revealed that he had an American Museum of Natural History grant, with some help from the Canadian government. He started doing cultural work, then switched to surveying and then ocean ice studies. Like Leffingwell, he was a man of many talents.) Little more was accomplished between his arrival on August 15, 1913 and July 1914, when Leffingwell left the Arctic for the last time.

Leffingwell's reason for leaving was unexplained and the departure seemed abrupt to say the least. Two men from the Canadian Arctic Expedition ships that had wintered to the east of Flaxman Island wanted to return to civilization, so they offered to help row the dory to Pt. Barrow. It was just over 200 straight-line miles, but took Leffingwell and the two men forty days because their way was blocked intermittently by shifting ice floes. With good sea conditions this would have been a five- or six-day trip. This effort reminds one of some of the rowing the Lewis and Clark oarsmen did on stretches of the Missouri River.

Although Leffingwell gives no hint as to when he worked on the Sadlerochit formation in the field, if it was about 1912 that would have been fifty-six years before the discovery of oil at Prudhoe Bay. He did not visit the oil seeps at Cape Simpson, but had reports of them and obtained a sample from C.D. Brower, the trader in Barrow, and had it analyzed. Later in life, Leffingwell expressed disappointment that his brief mention on the oil seepages in his *Professional Paper 109* caused much greater discussion than his very substantive documentation of extensive geology and geography. In the course of his work he named seventy-two topographic and physiographic features, as well as streams and landmarks.

Two brief entries in his USGS Professional Paper summarizing his exploration of northern Alaska show what a keen observer and documenter he was.

Figure 5.1 shows Leffingwell was an artist. He must not have been satisfied with the photograph he took of a huge boulder, dropped on Flaxman Island by a long-since departed glacier, because it was too dark. That didn't bother Leffingwell; he rendered it in a line drawing or pen and ink sketch, which you see here. I have seen the photograph in his collection, and the sketch is better!

An unusual item of note occurred when he was camped for a few days in April 1908 near the copious Shublik Springs, from which the water vapor froze to ice crystals. The weather was clear and calm, and the sun shining through the ice crystals formed a "parhelion" almost continually. The line drawing (Figure 6) is what he saw. The common term for this phenomenon is Sun Dog. He described the colors by assigning colors to different numbers.

Figure 5.1: Leffingwell's boulder sketch. Notice the binoculars for scale.

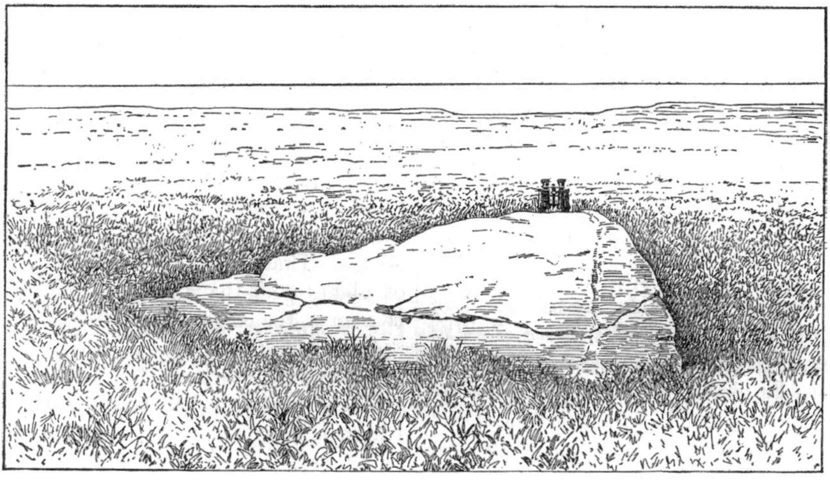

As noted, Leffingwell was so broadly educated that he was his own geographer, cartographer, explorer, astronomer, hydrographer, meteorologist, topographer, artist and zoologist. Leffingwell was obviously a person with many accomplishments.

Charles G. "Gil" Mull, now retired from many years as a geologist for the State of Alaska, and long an admirer of Leffingwell, helped to honor Leffingwell's memory for his contributions to Alaska's geologic and scientific knowledge in 1971 by obtaining National Historic Site recognition for the remains of his headquarters on Flaxman Island.

From Gil Mull in 2005, "His camp site is now a cabin that had been extensively modified by Eskimos over the years, but there are still brass ships fittings and woodwork that had to have come from the *Duchess of Bedford*...It was in fair shape [the last time Mr. Mull saw it], although obviously extensively modified from what Leffingwell left."

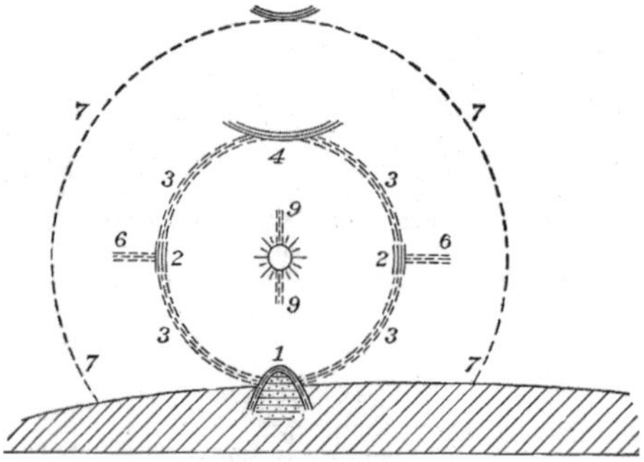

Figure 6: Leffingwell, not having the capability of color photography, used letters to identify different colors. A modern-day rendition of a Parhelion is shown in the color photo

Leffingwell estimated that he had pitched camp about 380 times, traveled by sled, foot, or small boat 4,500 miles, made ten local shipboard trips of 2,500 miles and sailed over 20,000 miles by steamship between Arctic Alaska and Puget Sound.

Setting an example was something that never entered Leffingwell's mind, but what an example he did set! His sole motivation seems to have been the desire to do something he believed needed doing. Although his work quietly steeped for years, others built on it a continuum of geologic knowledge that ultimately led to the discovery at Prudhoe Bay.

1

2

3

4

5

6

7

8

9

10

11

12

13

14

15

16

17

18

19

20

21

22

Chapter Two Photo Captions

1. *Duchess* locked in ice, eventually to be crushed. Leffingwell USGS124.
2. *Duchess of Bedford,* Ejnar Mikklelsen's and Ernest deK. Leffingwell's transportation to Arctic Alaska, in happier times, c. 1906. Leffingwell USGS260.
3. Leffingwell's cache in the summer. Note the smaller pelts, mostly foxes probably, and a couple polar bear hides. Leffingwell USGS145.
4. Winter camp of snow blocks (Eskimos didn't have them, but Leffingwell did). Note the stove pipe. What did he use for fuel? Leffingwell USGS147.
5. Spring season camp, Hulahula River. Leffingwell USGS91.
6. Land travel, backpacks on men and dogs. Leffingwell USGS140.
7. Emergency travel by ice floe on newly open water "lane." Leffingwell USGS27.
8. Leffingwell's headquarters with storage sheds to left. Leffingwell USGS231.
9. Arctic Ocean, crossing pressure ridge of old ice. Leffingwell USGS142.
10. Leffingwell summer camp on the Canning River. Leffingwell USGS164.
11. Tough sledding through more pressure ridges in new ice. Leffingwell USGS21.
12. Emergency travel by improvised raft going to mainland to retrieve boats lost from Flaxman Island in gale. Leffingwell USGS134.
13. A better view of a pressure ridge. If the man is six feet tall, the ridge is twenty or more feet high. Leffingwell USGS5.
14. Umiak, an Eskimo boat rigged for a sail, near Barrow, Alaska. Leffingwell USGS98.
15. Eskimo summer camp. Note meat drying on the wooden drift wood rack. Leffingwell USGS94.
16. Ernest deKoven Leffingwell, December 1961, 87 years of age. Photo from *My Polar Explorations 1901–14* (Leffingwell). Courtesy The Explorers Club, NYC.
17. Leffingwell on a waterfowl hunting trip at Flaxman Island, Alaska. Photo from *My Polar Explorations 1902–14* (Leffingwell). Courtesy The Explorers Club, NYC.
18. The most reliable way Leffingwell and Mikkelsen found to get across open water during their trip on the Arctic ice was to swathe a sled with canvas and use it as a boat. From *My Polar Explorations 1902–14* (Leffingwell). Courtesy The Explorers Club, NYC.
19. Winter Eskimo camp, with unfinished tent frame between two finished tents. Leffingwell USGS118.
20. They have their tails in the proverbial crack of ice. They look to be contented. Leffingwell USGS170.
21. If clothing or other gear gets wet, it must be dried. Leffingwell USGS150.
22. Winter camp of hunting party with five mountain sheep. Canning River in a fairly tall stand of willow brush, lots of fuel. Leffingwell USGS103.

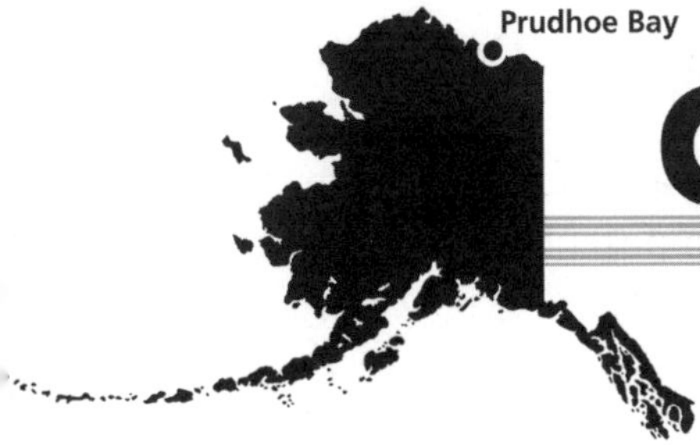

Chapter 3

The First Oil Activity: 1915 to 1923

In 1915 in the village of Barrow, Alaska, a government schoolteacher — also charged with medical duties, supervising the reindeer herds, acting as justice of the peace, postmaster, and so on — traveled with a friend from Wainwright to Cape Simpson, sixty miles southeast of Barrow (Figures 2 and 7). There they staked claims, in accordance with federal regulations, on oil seepages. The teacher was T.L. Richardson; his partners were William. B. Van Valen of Wainwright, Olaf Hansen, a Barrow trader, and Bert Panigeo and Egowa, two Eskimos.

The Eskimos, knowing of the oil seepages, had told Richardson about them. When they got to the Simpson area the seepages were easy to find because they were flowing from two large mounds over fifty feet high and about 200 feet in diameter. Van Valen related that they also could smell the oil for at least a mile before they got to the site. The visual detection of the oil was obvious because natural distillation caused it to separate into three or four different petroleum forms with different colors. These different fractions floated on water with an iridescent sheen of all the colors of the rainbow, making a brilliant display in the bright Arctic sun.

Richardson and Van Valen staked two twenty-acre claims, one on each seepage, and marked the corners of the claims with wood posts they brought for the witness stakes. On May 20, 1915, they placed the prescribed location notices — which included a description, measurements and location of each claim — in a tin can, and fixed it to the top of each post. They named their venture the Arctic Rim Mineral Oil Claims. There is no evidence anything was ever done with the claims.

Later in 1915, Alexander Malcolm "Sandy" Smith and W.H. Berry, gold prospectors, were on their way to Pt. Barrow after searching for a lead to a gold prospect given to them by an Eskimo. The gold failed to materialize, but Smith literally stumbled up to his knees into a pond containing the oily residue from the seepages. Smith was renowned as a prospecting, exploring, fur-dealing pioneer. He was a Scot who adopted and loved Alaska, and is rumored to have allowed that he would sooner be broke in Alaska than rich anywhere else in he world.

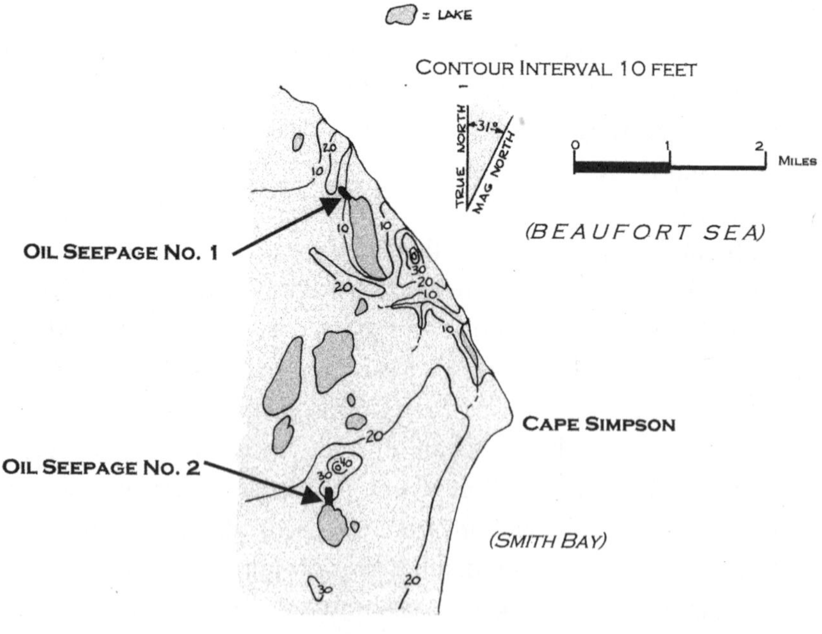

Figure 7: Cape Simpson oil seepages stimulated oil exploration interest. The seepages have accumulated for untold centuries. They are conspicuous, although the elevations are modest, on an otherwise featureless coastal plain. Philip S. Smith and J.B. Mertie, 1930, USGS.

With his lead dog, Slim, his indomitable will and high-minded spirit, Sandy was widely admired. His experience, extensive knowledge of the trails and dog sledding abilities led to him being orally crowned The Ace of the Trail. Smith went back to San Francisco after seeing the seepages and, with his entrepreneurial glands working, in the winter of 1920–21 assembled a group of capitalists who agreed to finance an expedition to file oil claims on the seepages. This group included several famous names of that era: King Gillette of razor fame; Henry A. Whitley, a Long Beach, California oil man; John Rossiter of the Pacific Mail Steamship Company; W.H. Crocker, San Francisco banker; and R.D. Adams, who made a fortune in the Nome gold fields.

The resulting group, referred to as the Adams Expedition, went to Cape Simpson in the summer of 1921 to investigate the seepages. R.D. Adams was listed as the chief, and the final report was sent to him. The geologist for the party was Harry A. Campbell who, at the end of his career, was credited with many oil discoveries in California. His assistant was Max Steineke, later to play a significant role in the Persian Gulf oil discovery. Campbell was the first geologist to survey the seepages. The Adams group staked forty-three 2,560-acre claims.

Campbell wrote the report for the party. He said in the report that Standard Oil Company had staked eight claims, presumably previously, and referred to these as a disputed area. He did not indicate the nature of the conflict. It is possible that Standard was there first, and the Adams people staked on the same ground. Such an occurrence was not unusual because additional recording of the claims was necessary for validation. Whoever gets to the Recorder first — the Federal Bureau of Land Management (in this case probably in Fairbanks) — to file a copy of the claim location and description, wins. The stakes at the claims of the Arctic Rim Mineral Oil claims of 1915 must have disappeared, or were not seen by the Adams group.

As would almost any geologist, Campbell recommended more surface work and opined that a core-drilling program might be worthwhile. In light of the teasing results of the subsequent core drilling in this area more than twenty years later during the Naval Petroleum Reserve No. 4 oil exploration program, it is fortunate the Adams group didn't pursue core drilling. The Navy drilled many tests and did find some producible oil, but not in commercial quantities. Geologically there was not much else for Campbell to report but the location and a description of what he saw. He collected samples for later analysis, which were 62 percent oil, 25 percent water and 13 percent a mixture of clay, vegetative matter and hair. The oil was composed of three different hydrocarbon fractions, much the same as were mentioned previously. There is no indication that any further activity was carried out on these claims by Adams or Standard of California.

The Navy's evolution from coal to oil resulted in the establishment of petroleum reserves in the United States. President Warren G. Harding signed Executive Order No. 3797-A that created the Naval Petroleum Reserve No. 4 (NPR4), which was established February 27, 1923, less than two years after the Adams and Standard of California claim filings. It covers 23 million acres of Arctic Alaska and is bounded by the continental divide in the Brooks Mountain Range on the south, the Arctic Ocean on the North, generally by the Colville River on the east and to a line parallel to, and just east of, the 162nd meridian on the west (Figure 2).

The significance of these seepages cannot be over dramatized. Their presence and the fact that valid claims had been filed on them set off alarms within the federal government, leading to the establishment of NPR4. Oil seepages are not always found over oil fields, but it has happened. An oil seepage is one of the most certain indicators that there is oil somewhere in the subsurface, and stirs the exploration juices in petroleum geologists. The path is usually torturous and the source is seldom, if ever, found, but in the early years of oil exploration, seepages motivated much oil exploratory activity. It cannot be overstated that absent the Cape Simpson seepages, the following events leading to discovery at Prudhoe Bay may never have occurred. The geological exploration knowledge in Arctic Alaska would have unfolded much more slowly without the foundation of the seepage knowledge.

The mindset of the early petroleum geologists who learned of the seepages at Cape Simpson probably led them to some exuberant speculating about their meaning, but almost certainly they did not believe that some forty-five years later a company named ARCO would discover the largest oil field on the North American continent. That would have been in the realm of fortunetellers, soothsayers, crystal ball gazers and clairvoyants. However, it certainly was another noteworthy event on the path to discovery at Prudhoe Bay. But first there would be a concerted examination of the geology and geography of NPR4.

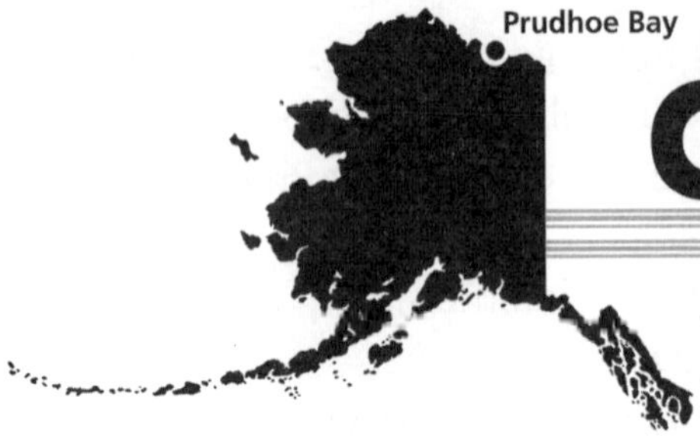

Chapter 4

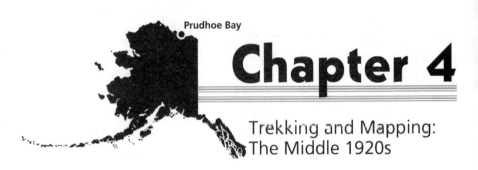

Trekking and Mapping: The Middle 1920s

The Bureau of Engineering in the U.S. Navy invited the United States Geological Survey to begin to explore and document the geography and geology of the recently established Naval Petroleum Reserve No. 4. (NPR4 is used throughout the text and on the maps. Later the designation was changed to NPRA, Naval Petroleum Reserve Alaska.) In the summer of 1923, the USGS sent a geological expedition to begin surveying the northwestern and northern coasts of Alaska in the NPR4 (Figure 8). Most of the work was done along waterways because that was the easiest mode of transportation. Sidney Paige, a geologist, was appointed leader of this expedition, which was divided into three field parties. The point of departure was Nome, Alaska. Two parties went to Wainwright, on the northwest coast of Alaska and 100 miles southwest of Pt. Barrow, to begin their work.

One of these parties, composed of Paige and E.C. Guerin, topographer, surveyed the northwest coast from Wainwright to Pt. Barrow, thence southeast to Cape Simpson, where the oil seepages were examined and mapped. Paige was exploring NPR4, so it is likely he speculated on the seepage's significance. Wouldn't it be interesting to know his thoughts in light of what was to follow?

The second party also worked from Wainwright twenty-five

miles northeast to Peard Bay, then portaged east to the Inaru River and traced it forty airline miles farther east to Dease Inlet, southeast of Pt. Barrow. James Gilluly, geologist, and James Whitaker, topographer, were in this party.

The third party, formed by William T. Foran, geologist, and Gerald FitzGerald, topographer, went to Cape Beaufort, also on the northwest coast of Alaska. They worked their way 150 miles northeast to Wainwright. In addition to mapping the coast, they ascended several different streams and mapped some of them for forty miles. They estimated their work threw light on the geology and physical features of about 10,000 square miles. This was the beginning of Bill Foran's long involvement as a USGS geologist with NPR4 over a period of twenty-five years. In Smith and Mertie's later, formal report, "Geology and mineral resources of northwestern Alaska," in USGS Bulletin 815, they had this to say about Foran's leadership of the 1924 trip, "[it] was an achievement that could have been carried through only by…the resourcefulness and insurmountable pluck and generalship of Foran the leader." Many years later, in 1970, John C. Reed, who was in charge of NPR4 exploration, gave Foran credit,

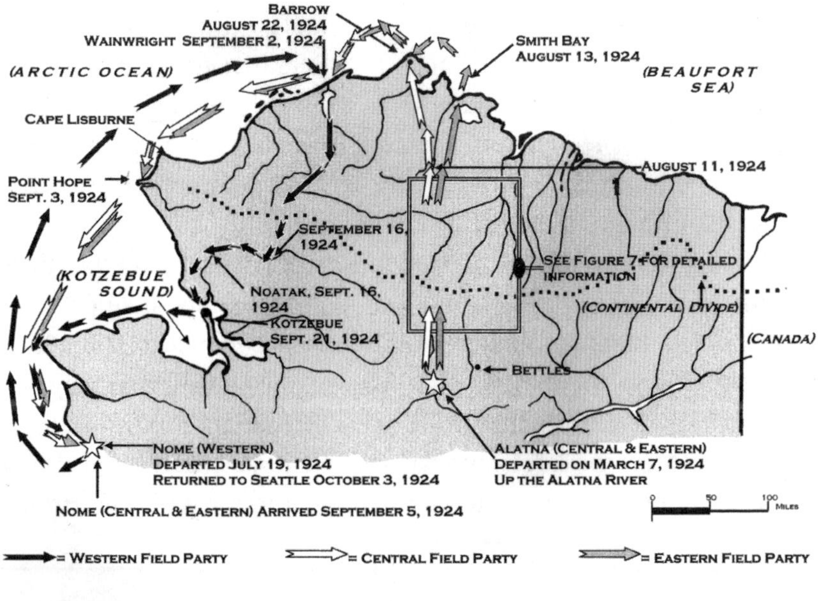

Figure 8: The first USGS exploration of Naval Petroleum Reserve 4, 1924–1927.

"He, as a Naval Reserve Lieutenant, was really responsible for the whole idea of the Pet-4 (NPR4) exploration starting in 1944." Simply put, Foran could always be counted upon.

Gerald FitzGerald, like Foran, was an active participant during the subsequent exploration of NPR4 in the 1940s and 1950s His name continually appears on Alaska field parties for the USGS. He ultimately became chief topographer of the USGS.

After the 1923 field season, it was apparent that exploration was needed in the interior of NPR4. If a full season's work was to be accomplished, the parties would have to be in place and ready to work when the snow was gone and the rivers navigable. This meant the field crews of geologists, topographers and camp hands would have to assemble supplies and equipment during the dead of winter, then travel to a position on a river and be ready to begin work when the river became passable.

Since Peters' and Schrader's expedition in 1901 the Alaska Railroad had been built, greatly facilitating the party's travel to the departure point in interior Alaska. This venture was led by Phillip S. Smith, topographer, with J.B. Mertie, Jr. as the leading geologist. They went to Nenana by rail, and thence 125 miles to Bergman by horse-drawn mail stage, negating the 1,100-plus mile sled dog route from Whitehorse that Peters and Schrader mushed. Even so, the Smith party endured many days of –50°F temperatures. They left Washington, DC, in mid-January 1924. It was forty-three days later, on February 27, that the party left Nenana on their trek over the Brooks Range to the Arctic coast.

During February 1924, they significantly depleted the Northern Commercial Company's inventory to supply the nine-man crew for close to six months. Winter clothing was contracted from local Natives because they knew what was most efficient, and they had years of experience making it. The crew members had to go as far as 150 miles to get the large number of dogs they needed for sledding. They hired local freighters to help them get to a pass in the Brooks

Range. To get there, special bobsleds had to be built to haul four Peterborough cedar freighting canoes. These were custom designed in decreasing sizes, so they could be nested in one another.

The party crossed the mountain pass on April 2, 1924, and then had to travel until April 19 to a lower elevation where there were enough trees to supply firewood for forty-five days. That was when they estimated the stream, which they hoped would be navigable, would thaw and they could begin their journey and survey the route.

While the group waited for the spring thaw, they worked out of their camp on ten-mile excursions to do topographic and geologic mapping. There was much boredom in the camp when weather prevented work. Reading helped, but there had not been room enough to bring adequate books to fill the voids. This tedium came to an end on May 30 when sufficient water to float the canoes began flowing. The impending resumption of their journey was announced several days prior, by the thunderclap-like roars of the ice breaking on the smaller streams. Their winter outfitting was discarded when they left the winter camp. Little did they know the dangers and physical exhaustion they were to encounter. Nor could they have imagined that their colleagues in the western party, led by Bill Foran, would come face to face with starvation.

They divided into two parties (Figure 9). Smith, Lynt, Dodge and Clark formed the central party, and J.B. Mertie, Jr. (he worked thirty years as a field geologist for the USGS), FitzGerald, Tait and Blankenship formed the eastern party. The plan called for the central party to descend the Killik, on which they had been camped. They were to move as rapidly as possible to the Colville River and then map upstream until they could find a suitable divide they hoped would lead them to the north-flowing Mead River.

Both parties had a difficult time even going downstream because of rapids with many high, standing waves where the streams crossed remnants of glacial moraines. It was also difficult navigating small channels flowing on top of ice. The ice could

Figure 9: Smith and Mertie winter camp for summer exploration, 1924. The central and eastern parties of the Smith and Mertie expedition stayed at their joint camp site on the Killik River from April 6 until May 30, 1924, when they began their geologic and topographic surveys after break-up began. They planned to travel different routes from the camp to the Arctic Ocean, but their lack of local geographic knowledge eventually led them both to the Ikpikpuk River where they met on August 8 and again traveled as a unit.

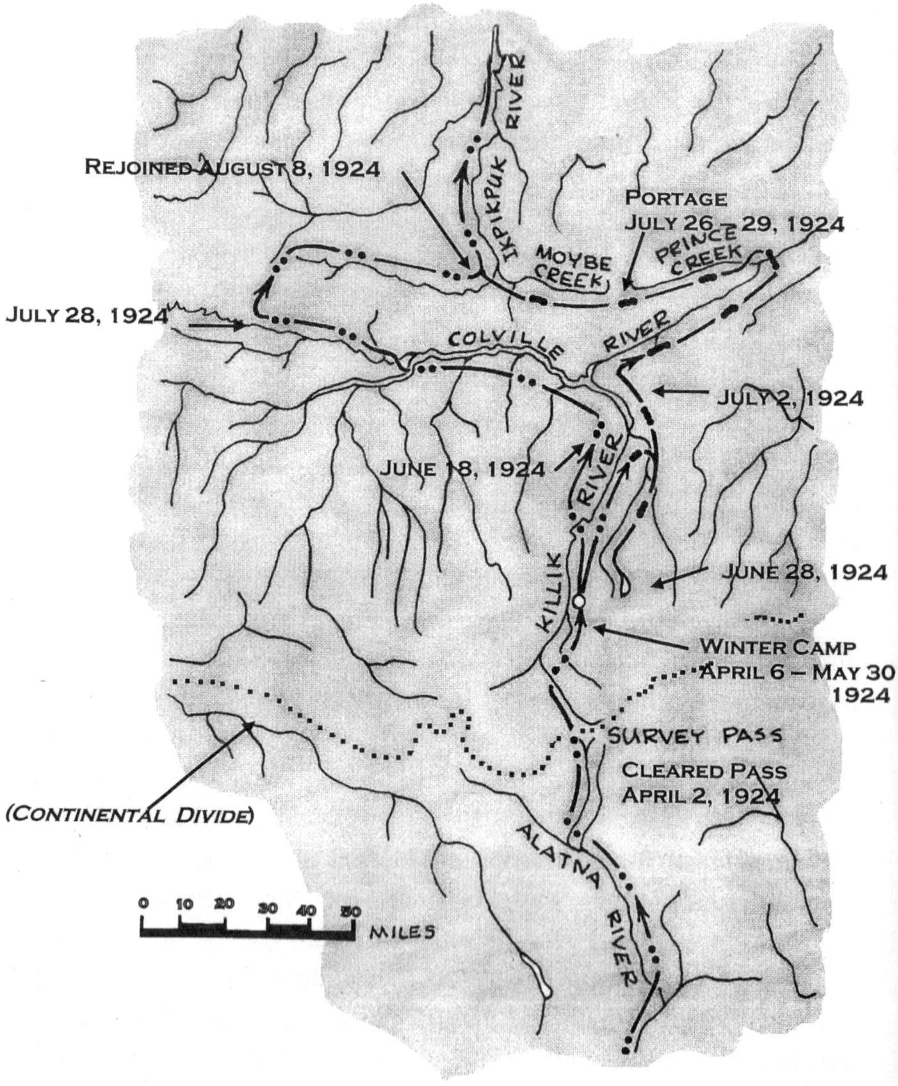

break loose from the bottom with no warning and begin floating, leaving them no control over their destiny. In addition, ice was also breaking loose from the stream sides. They were in danger of being struck when unavoidably close to a falling mass of ice, or swamped by a wave induced by an icefall. Conversely, at other places the channels were braided with too little water to float the canoes.

Here is Mertie's account of one experience:

> We came to a stretch of river which looked possible to run in the boats. Fitz and I looked at it from the top of the bench [a hill]...and all thought we could make it. So we pulled into it, Fitz and Tait leading in their canoe. They got through, but Blank and I, in our heavier canoe, floundered. *[sic]* We didn't hit any rocks, but the big waves in the rapids just combed over us and filled the boat with water. We submarined the last rapid holding onto the boat without it turning over, and managed to get ashore with the canoe and cargo, just above one of the worst rapids. It would have been all off if we had gotten into that one...I decided then and there that we would run no more rapids that summer.

The eastern party was to map the geology of the Killik to the Colville, then descend it until they could find a suitable drainage that would take them to the north-flowing Ikpikpuk River, which more or less paralleled the Mead on the east.

Both crews found the divides, but it was a difficult undertaking. The respective streams became so small at their headwaters that the men would have to enlarge the channels by digging where they could with shovels. When this became futile, the canoes were pushed and dragged from one small pond to another. Finally, the only alternative was to portage the canoes and all their supplies. In preparation for this, one of the crews sent two men to scout the situation. They made an eighty-six-mile round trip in three days to make certain there was a stream to descend on the other side of the divide. Mertie comments again:

> We got back to our spike (temporary advance) camp about 2 A.M. We were nearly all in but happy and relieved that a portage to the Ikpikpuk had been found. It had rained most of the night and we were wet from head to foot when we finally got to bed. I lay on my back and chattered my teeth all night while the rain beat down on the tarp, gradually soaking through and wetting the bed. What a night! We had traveled three days, two of them with packs on our backs. I hoped never to put in three such days again.

A short few days later, he allowed as how he was tired, "and so was my poor old broken foot." Like Lewis and Clark, how did any of them do the things they did?

Then began the physically daunting task. Four men had to drag the canoes, one at a time, a distance of eight miles, all in one day. Next they backpacked all the equipment. This was done in seventy-pound loads. Those packs must have felt like they weighed twice that at the end of the trip. Remember, too, this was through swampy, uneven tundra. There was no trail except the one eventually made by their repeated trips. This took eight days, and they were near exhaustion many times. Mertie recalled how difficult it was, "seventy pound to the load [backpacks] and two round trips a day." (Thirty-two miles total.)

The men were appalled when they saw that the stream they were to descend was also dry. This could only mean still another ordeal before reaching navigable water. Some prayed. Some simply waited. Others despaired. But thankfully, some days later, Providence intervened. It began to rain, a little at first, then harder. Within forty-eight anxiety-ridden hours, the stream was bank full. The men were ready to move forward.

Geographically they were in a "no man's land," so it came as no surprise that the central party had chosen the wrong divide and both parties descended the Ikpikpuk River. Neither group knew this until the eastern party saw signs that the central party was ahead. The eastern group moved with more dispatch and soon rejoined the central party on August 8, and they jointly surveyed downstream.

Meanwhile the western party under geologist William T. Foran with O.L. Wix, topographer; H. Lonseth, rodman; H.G. Hughes, boatmen; and F.W. Belgard, cook, left Nome for Wainwright on the northwestern coast of Alaska from which they surveyed this part of the NPR4. This crew also had a difficult time because of the absence of geographic and topographic information. This crew surveyed inland from the northwest coast of Alaska with the objective of crossing a divide in the Brooks Range and returning to Kotzebue. This village is located on Kotzebue Sound, which is a large arm of Chukchi Sea on the northern west coast of Alaska. Kotzebue is located roughly a quarter to a third of the way from Nome to Pt. Barrow.

Once on the south side of the divide, they needed to find a south-then west-flowing stream that eventually would lead to Kotzebue. They had lost so much time that when they did find the right drainage it was September and freezing had limited the flow of water. Here is Foran's account as to why they lost time, " a six-mile portage. Without freshet portage would have been 25 miles. Even a six-mile portage of two canoes and equipment was some task." This occurred August 16–19, 1924. By August 23 they had dragged the canoes so much they had worn holes in the bottoms. The weather varied from cold with snow to clear and hot. Their extreme effort caused dissension. Belgard said he hired out as a cook, not a packhorse. They decided to cache over a ton of supplies and equipment and one canoe, and had to start backpacking. Attitudes improved, but not for long as their situation continued to deteriorate. They had passed the point of no return for the cache of food they had left behind. Ptarmigan continued to be their mainstay, but they were few and far between. Foran's men were obsessed with thoughts of food, and became insolent to the extent he thought mutiny was imminent; this was September 1, 1924. He recorded [September 1st] “Sounds late to be in this neck of the woods with only a summer outfit.” Indeed it was, and the weather recorded was cold, cloudy and snowing. [The field notes by Bill Foran make for interesting reading. The most interesting parts are transcribed in the Appendix for the convenience of those wanting to read them.]

The food situation had become so acute, Foran dispatched two team members to return to the cache for supplies, in spite of the fact they had gone beyond what they thought was the point of no return. Foran and a man named Wix, the topographer, went ahead, looking for the Noatak River or its tributaries. It flows generally east to west and is the first major drainage system at the foot of the Brooks Mountain Range on the south side. At 5:00 p.m. Wix began his return to camp, but Foran kept searching because he had not found the route to the promised land of Kotzebue. He gave up at 10:00 p.m., tired and hungry, having had nothing to eat but a biscuit and sardines at lunch time. He walked all night, not recognizing landmarks because of heavy frost and a certain inability to function adequately in his poor physical condition. At 8:00 a.m. the next day, having been on his feet twenty-four hours, he finally found himself in the same valley as their camp, but three miles east of it. It took him another hour and a half to reach the camp where they were preparing to send out a search party. His feet were in terrible shape because he had been hiking in a new pair of leather doughboy shoes. (A doughboy was the World War I equivalent of the World War II GI, or Dogface.) Foran ate some corn and drank some cocoa and got in his sleeping bag. He woke briefly at 10:00 p.m., ate again and slept until the next morning.

The party continued to move south, certain of the direction, but with no idea as to whether they were on the right drainage system or whether they would end in a box canyon only to have to search for another route. They found themselves in an area where there was absolutely no wood for a fire, and so added perpetual cold to their misery. It hardly mattered now that there was no food and no fuel with which to cook it. It was now September 9, and Foran noted in his journal, "we are not [underscored heavily] in a climate suited to summer clothing." The weather recorded was "very cold and cloudy." At this low point of morale they crossed the high point of their trek. Leader Foran described it in understandable terms, "Like dropping out of northern Alaska into California." As a group they believed that freezing to death was a distinct possibility, but now being on the south slope, that eventuality virtually disappeared.

At this point they had to cache items that were their stock in trade, and left them only as a matter of last resort: survival. They left the tools, surveying instruments and rock hammers. They left specimens of fossils and rocks, which they needed to augment the interpretations of the rock outcrops they had seen and described. (That is, the stratigraphy and stratigraphic sequences defined in the Leffingwell chapter.) They also left personal items like towels, undershirts and leather shoes. They were on streams going down current, but the problem was lack of water because the low temperatures in the mountains were freezing the water at its source. Little by little they got lower, to a point of the drainage system where source water was not being frozen. Travel eased, but hunger still ruled and was causing great dissension. Their strength had deteriorated to the point that they only had enough to function two or three hours without rest. If the river became too rough they would have to stop until they regained as much strength as they could muster before going on.

The "sea gull incident" happened on September 17, seven days after they crossed the divide. It was a mixed blessing, or in today's vernacular, good news–bad news. A lone sea gull came within range of Foran's gun, and he described it as a "lifesaver and tasted like chicken." The men were so ravenous every morsel counted, and all but Foran accused the cook of wolfing down the gizzard. The argument got so heated there was the threat of "hot lead to assist digestion of gizzard." Foran assigned one of the least belligerent to sit on the gun and ammunition.

They continued, making good time now — seventy miles in one day — and a day later came to an Eskimo camp where they were well fed. One of the men was still arguing with the cook about the gizzard. The following day they arrived at a Mission at 11:00 a.m. and ate for eleven hours. Eating in moderation after a starvation diet is the medical rule, but who can restrain themselves? They ate so ravenously that all of them had a severe bout with diarrhea. For some, the problem persisted for two weeks.

Kotzebue came into sight early on September 20. In spite of being safe and well fed, on September 22 and again on September 30, there were heated arguments over the gizzard. The argument had persisted for thirteen days until the trader, hearing the arguments, told them that sea gulls have no gizzard. Bill Foran concluded his log with the comment, "Exciting trip from start to finish."

This experience enabled the same crews to be more efficient in the subsequent years of 1925 and 1926, and added more geologic and geographic information to the accumulating information.

In the USGS Bulletin 915 (for the years 1923–25), P.S. Smith and J.B. Mertie, Jr. had a section in which they discussed briefly the petroleum potential. They were bold, but not irrational, to make predictions on no more data than the oil seepages at Cape Simpson and oil shales they found. The oil shale samples were found in rock outcrops in only one locality. The geologic age of the oil shale sample outcrop could not be determined because not enough definitive rocks were exposed. They believed the geologic age of the shales to be from the Lower Cretaceous period. An analysis of the shales "indicate an oil content of over 50 gallons to the ton of rocks, so that a bed of shale 1 foot thick contains about a millions tons of rock for each square mile of its extent." This means there would be well over a million barrels, a significant number. Here is how to put a million barrels in perspective. If the oil shale rock was 100 feet thick, this would mean one square mile would contain more than 100 million barrels of oil, and an oil field this size is considered to be a major field.

Unfortunately, oil moves extremely slowly through shale rock, so it is not a good reservoir. Reservoir rock must have two qualities. It must be porous (have voids like a sponge) so it has space within it to hold oil, and it must be permeable so the oil is free to move through the rock from pore to pore (void to void), and be produced when penetrated by a well. If there is sandstone (a reservoir rock) within oil shale (a source rock), over unknown years of geologic time the oil will slowly move from the shale source rock and fill the sandstone reservoir rock.

Smith and Mertie said the petroleum potential was doubtful in the older and deeper rocks, but they did find some evidence of oil shale in more geologically recent rocks. During the development of Prudhoe Bay some forty-odd years later, those more recent rocks, when found in the subsurface, were believed to be the source rock of the Prudhoe Bay accumulation.

These comments about their predictions are not meant to ridicule, but to compliment, because they had so little information with which to work. With only surface outcrops on which to make predictions they had no way to postulate what might be found geologically underground. These last few paragraphs point to how much in the dark all oil explorationists work until more complete data becomes available. These data become available because someone has an idea, and that individual, or more likely the company for whom he or she works, has the money and is willing to gather more data and to eventually drill to test the idea. Even then, many great fields are drilled for one reason and oil is discovered because of some unexpected geological occurrence that could not be foretold before the well was drilled. It is quite remarkable that one of three predictions was correct, in light of Smith and Mertie's meager knowledge.

This work in the middle 1920s was another building block that eventually led to the Prudhoe Bay discovery. It would turn out that

William Foran, a proven leader of these years, would continue his leadership in the future in the exploration of NPR4.

The activity that has been discussed to this point might seem continual because it has been linked in this narrative. In the real world there were so few people working so vast an area, with primitive transportation, that much remained to be done. In spite of this, the period from 1926 until 1944 was the longest devoid of any geological exploration in the area. There was nothing to motivate more detailed geological exploration. The next significant activity occurred in 1944 when World War II and the need for fuel for ships became the catalyst for an unprecedented effort to evaluate the petroleum potential of Naval Petroleum Reserve Number 4, on the journey to discovery at Prudhoe Bay.

3

4

5

6

7

8

9

10

11

12

13

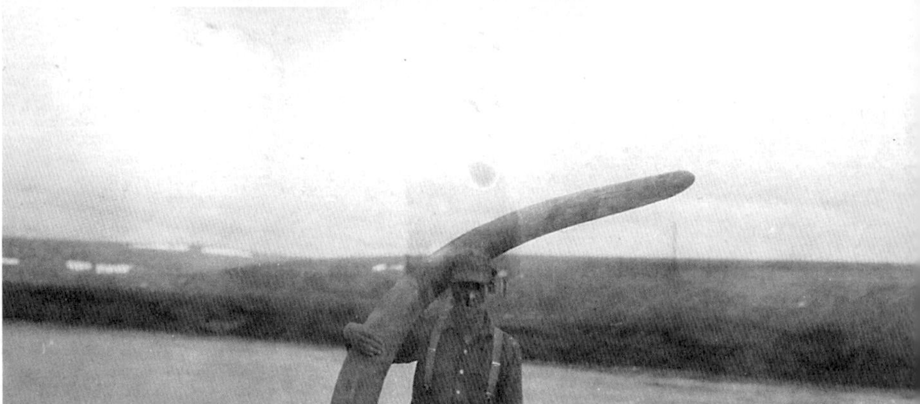

14

Chapter Four Photo Captions

1. Arrived at Alatna, February 17, 1924 via steamship, Alaska RR, USPS Mail Stage (horse-drawn) and now by dog sleds over the Continental Divide to winter camp on the Killik River.
   Smith USGS1499.

2. "Spruce" camp, March 26. Note large canoe crate.
   Smith USGS1520.

3. Mertie with sled, April 21. Mertie USGS1125.

4. Bush barbershop, April 22. Smith USGS1545.

5. Uncrating the canoes, May 23. Mertie USGS1129.

6. Mountain scene near winter camp (also Camp 22), May 23. Mertie USGS1132.

7. Winter camp, May 24. Smith USGS1571.

8. A final view of winter camp and the all-important cook tent on June 1. Smith 1600 USGS

9. Running the rapids, central party, June 8. Smith USGS1628.

10. Central party camp, June 10, drying clothes on brush. Smith USGS1642.

11. Eastern party, June 12. Mertie USGS1180.

12. Eastern party on the Colville. Mertie USGS1185.

13. Mastodon tusk, central party, July 4. Mertie USGS1681.

14. Parties recombined, Chipp River, August 11. Mertie USGS1213.

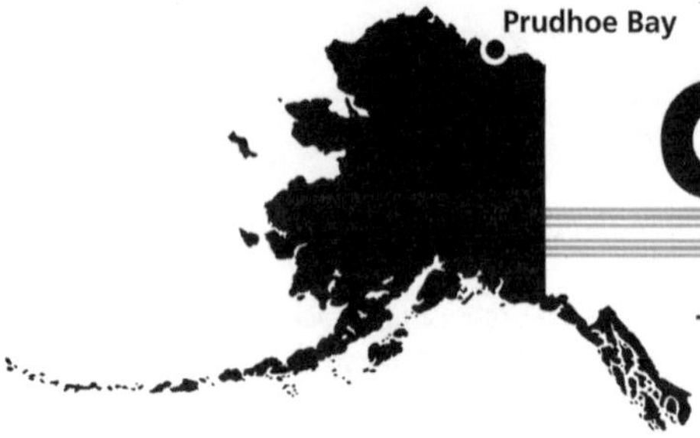

Chapter 5

Turning Point: NPR4

"World War II was in the making, and international unrest was growing. Finally, along with most of the rest of the world, the United States was plunged into the conflict both in Europe and in the Pacific. This was a different type of war, mechanized beyond previous imagination, and requiring almost unbelievable quantities of petroleum products…and the shortening of global distances as better, faster, longer range aircraft were developed. The whole pattern was such [that] there was need for a more complete knowledge of the petroleum potentialities of Naval Petroleum Reserve No. 4." These are the words of John C. Reed, Commander, USNR in his *History of the Exploration of NPR4*.

Since the last year of geological exploration in 1926, there had been no further exploration of NPR4 until 1944. The Smith and Mertie excursions in 1924–26 had concluded that rocks of Cretaceous geologic age were probably the best objectives. This was a good presumption because rocks of this age have produced much oil in the Prudhoe Bay area since the original discovery.

Frequently an insignificant event or a person in a position of relatively modest importance can stimulate activity of great magnitude. William T. "Bill" Foran, the savior of his field party in the western part of NPR4 in 1923, was Lt. Foran, USN, in 1943. He prepared a recommendation for the Navy to evaluate the petroleum potential of NPR4. This recommendation filtered up the chain of command until the Cabinet Departments of Interior, War and Navy and President Franklin D. Roosevelt were involved. Foran and others with experience in shipping, building and oil field drilling were sent on a reconnaissance mission to northern Alaska. In retrospect, this was window dressing because one wonders what of real importance they could have learned in a month's time. They saw everything as doable and said that the oil seepages at Cape Simpson were reason enough to go ahead, no surprise there. The recommendation again went through channels to President Roosevelt who gave his approval, and this began nine years of exploration with what was, or close to, a blank check from the U.S. Congress.

The Chief of the Bureau of Yards and Docks sensed that approval would be forthcoming. Accordingly, he had already arranged for supplies and logistical support and a flotilla of ships with 8,448 tons of supplies, equipment, Seabees (WW II Navy constructors) and stevedores to unload the ships that left Tacoma, Washington, in July 1944. The flotilla disgorged its cargo in late August and September on the beach near Barrow, Alaska. The next year the supplies brought north amounted to 17,000 tons. These annual resupply missions continued until the end of the program in 1953. The window in the location of ice usually occurred in only a few weeks near the end of August and early September. Each year there was apprehension, and there were some tight situations when the ice moved in unexpected and unwanted ways, but nothing completely thwarted the operation. ARCO and British Petroleum had to deal with the same annual uncertainties after Prudhoe Bay was discovered, but the time window was much shorter and the risks were greater because they had to go another 200 miles east where the ice situation became increasingly worse. Tens of thousands of fifty-five-gallon oil drums were a very large shipping, handling and storage problem. A big breakthrough came when the Navy erected storage tanks and began to ship and store fuel in bulk quantities.

Through the years the Navy sent supplies on tractor trains to all the different drilling locations. There was so much drilling in the Umiat area, just west of the bend in the Colville River where its course changes from northeasterly to north, that it became a sub-depot from which other areas were supplied. Tractor trains, consisting of up to a dozen sleds pulled by a large crawler tractor, serviced these drilling locations and Umiat. There was a sled that carried living quarters and a mess hall. A round trip from Barrow to Umiat was 635 miles and required sixteen days. One year these trains moved almost 6,000 tons of equipment over 3,711 miles of trails. Radios were unreliable, so the trains were serviced daily by bushplanes delivering mail and incidentals and for emergencies.

Umiat became an important base for oil industry activity beginning in the late 1950s because the Navy had established so much relatively permanent infrastructure there. It was so functional that at one time the latrine was a huge twelve-holer. At least one major oil company president was not charmed by the stimuli to his olfactory nerves, the absence of privacy or the level of ambiance.

In addition to the drilling the Navy planned for the program, they enlisted the United States Geological Survey to conduct geological studies to help guide and interpret the drilling. It was the geologists' job to find the rock outcrops, the places where forces in the earth and erosion had exposed rocks. At the outcrops, the geologists' work was primarily twofold: they studied the stratigraphy and stratigraphic sequence of the rocks so they could correlate similar rocks wherever they were found. They also surveyed the attitude (the strike and dip) of the rocks to determine existence of anticlines and their magnitude (their length, width and amount of closure), which suggest they are near or on part of an anticline, and thus justify more evaluation. As previously noted, stratigraphy is a description of the composition, texture, grain or crystal size, fossil content, geophysical and geochemical properties, and thickness, sequence and correlation from one place to another. Stratigraphic sequence is a succession of sedimentary rock beds chronologically arranged from older strata on the bottom upward through increasingly younger strata. Attitude is the deviation of a rock formation surface from horizontal. Rocks change in character, sometimes slowly, sometimes rapidly. Learning how and where they change is important. Many of these changes can be used to help interpret what is happening between outcrops and, to a lesser degree, in the subsurface between wells.

Another activity the geologists did was to survey the ground to map surface anticlines. There was one at Umiat. (Figure 2) An anticline is the name given to an area where powerful forces in the earth have forced horizontal layers of rock into an arch. Geologists took measurements that made it possible for them to draw contours on the surface of the arched rocks. A contour is a line that is the same elevation, or distance above sea level, everywhere along that line. The distance from the lowest contour that goes all the way around the anticline to the highest elevation of the anticline is called the closure. Foran and Woodward found the Umiat anticline had 750 feet of closure. This is significantly large. In Wyoming in the early days it might have been referred to as a sheepherder anticline because it would have been so obvious even a sheepherder could see it.

The geological field parties were in the “field” in Alaska from about mid-May until mid-September each year. Temperatures were generally between 30°F and 70°F, but it could snow any given day. There were typically four parties in the field each summer, with at least two geologists, a camp hand and cook in each party. Over the ten-year period the geologists probably worked about 9,600 man-days. Charles R. Metzger documented the hardships they experienced in his book, *The Silent River*. Metzger chronicles one of the first summer’s fieldwork of a party led by George Gryc, who was a part of the NPR4 exploration team for all the years of that effort, and thereafter was one of the "deans" of the geology of northern Alaska.

Just getting to the outcrops is hard, physical work. Metzger's brief analogy paints the picture.

> Tundra is miserable stuff to walk on. You are lucky if on foot you can average a mile and a half an hour over it. If you were to take a choppy sea surface, instant freeze it, then cover it with a layer of loose earth, mud, water, grasses, mosses, dwarfed junipers, heathers, etc., you would have something like tundra."

The tops of tundra always seemed too far apart, and the valley too close, to make a consistent pace possible.

> In Arctic Alaska nearly everything tastes or smells like tundra; the air, the water, the game. Gryc mentioned that his wife said he smelled like tundra for six weeks after he returned from the Colville. The old Army parka that I brought back from the Arctic continued to smell like tundra until it finally wore out. The moose that George Hipple later shot, having dined on lily pads, tasted less like tundra and more like what we down south know as veal. The mountain sheep we ate, having grazed on moss and lichens growing far above the tundra, tasted like young goat or like something between veal and pork.

The tundra was also home to clouds of mosquitoes, which required the men to wear head nets to prevent them from inhaling the insects.

The camps were staged so the crewmembers could travel by boat from one campsite to the next. This was treacherous, as can be understood from the following passage from *The Silent River*.

> At one particularly difficult place, so difficult it took four or five of us to manage a single boat, the current tore Banks away

> from our boat and swept him down stream. Stef, who was at the stern of our boat downstream from Banks, opened his eyes as wide as I have ever seen anybody do, then immediately reached out and grabbed Banks by his parka hood as he swept by on his way toward real trouble.
>
> Banks, completely soaked except for his head and watch cap, had immediately to change into warm dry clothing, even though he was wet up to his crotch again only in a few moments after rejoining us at work in the white water. It was snowing lightly at the time.
>
> After a couple of weeks, a great deal of effort, and some excitement, we finally got to the end of the most difficult rapids. Then a huge boulder in the middle of the river sent up a wave of roaring white water in a place where we could not get around it. The channel meanwhile was cut about fifteen feet deep into the surrounding soil, and so all we could do was unload our boats, carry everything up from the river to the tundra above, including our boats, and re-launch them on the downstream side of this obstruction. We had been wading in cold water sometimes clear up to our shoulders twelve to eighteen hours a day.

One of the first campsites of the Gryc crew was near an Eskimo camp. There became a mutual bond between the two camps. The geologists learned much about Eskimo culture and saw many examples of their cognitive abilities, not to mention hand–eye dexterity, as this description of the cat's cradle game shows.

> With fingers as nimble as any surgeon could ever hope to develop, he portrayed for us, using the string, symbolic pictures, both still and in motion. First he constructed with the string looped over his fingers the standard elementary abstractions well

> known to anyone who has served time as a kid in the Pacific Northwest. Then he progressed to making abstract portraits and landscapes in string. Then he moved onto more difficult figures, a weir for example, which required him to use his teeth as well as his fingers in order to stretch the string into a complex pattern suggesting a taut dip net. Finally he moved into mobile figures, one of the more interesting portraying, as he said, "Eskimo chasing caribou," during the performance of which one loose knot chased another across the string screen stretched between the carefully manipulating fingers of his two extended hands. Finally he produced his masterpiece, "Mouth eating" involving a complicated knot loosely and flexibly extending and contracting so that the loops of the string simulated the lips of a mouth chewing. Dinner being long since ready, and we having been thoroughly well entertained, it finally occurred to us that the man wanted to be fed. So we fed him. He trotted off toward his home on snowshoes, well fed, immediately after dinner.

This established a precedent; from then on the Eskimo leaders would frequently find "important" reasons to visit the camp about that time of day.

The field crews depended to a great extent on game for a significant part of their diet. The rest of the food, mostly canned or dehydrated, was cached each spring well in advance of the field season. Marvin Mangus was assigned this responsibility fairly early in the program and did it from then on. Before the actual caching operation in the field, the rations had to be segregated and packed in steel fifty-five-gallon barrels. Years later Marvin went to work in the oil industry and eventually was the senior well-site geologist who supervised the evaluation of the discovery well

Bears were the bane of many of the geological field parties, as Marv relates in the following story.

> A bear is nature's greatest locksmith; he can open anything. On one occasion when the crimping tools wouldn't work, the mechanics secured the tops with nuts and bolts. The bears opened every drum. A lot of the food inside was in cans, but the bears were equipped with their own can openers, strong jaws and sharp teeth. Each can was smashed steamroller flat but the teeth marks told what had happened to the contents.

> Bears would save themselves a lot of trouble if they'd only learn to read the labels; they'll go to just as much trouble with cylindrical objects like flashlights or cooking stoves, as they will with a can of peaches. The only thing that seems to be safe from them is dried fruit. One time a family of bears got into a 700-pound cache and disposed of everything, including opening and draining Coke bottles, but didn't touch a nice big package of dried apples.

It would seem that the geologists and bears had an ongoing love/hate relationship. The bears loved the caches; the geologists hated losing them.

The term "bush" is much used in Alaska. It does not refer to a form of plant life. It is ill defined, but generally means anyplace one cannot go to by road, and if the road is poor enough even then the area is referred to as being in the bush. If you have to hike, ski, fly, boat, dogsled or snow machine to a destination, it is definitely in the bush. Size is not part of the equation. Kotzebue is considered to be in the bush and it has a population of more than 2,500. All the pilots that serve such areas are referred to as bush pilots.

Bush pilots flying bushplanes did much of the ferrying of supplies and personnel during NPR4 exploration. Any single-engine plane might be used for bush service. The ones that fit the bill best had STOL capability — short take off (and) landing. Depending on the season and locality, the planes might be equipped with skis, pontoons (more commonly referred to as floats) or wheels if there was the semblance of an airstrip, frozen lake or a river bar. Bushplanes delivered all of the spring caching, then later, mail and resupply. Bush pilots were very resourceful and jacks of all trades; they didn't last long if they were not good. Here are a couple of quotes from *The Silent River* to put this form of travel into perspective.

> Waino would land and take off from impossible places. That, of course was his business and he did it well, which was why he was still alive. Waino circled our camp and the stretch of river nearby about five times, looking it over very carefully, before he came in to land…he would land damn near anywhere if you gave him time. He came in cross wind on a curving stretch of our river and touched down in white water [rapids], calculating to remain up on the steps of his floats until just before the floats settled down into the slack water just above the rapids. He coasted into the shore, with his engine off…just in time for us to grab his floats…and pull him aground. He stepped out from the cockpit onto one of the pontoons looking just like a bush pilot, improbable. I have yet to see a bush pilot that looked anything like the impeccably tailored, clean cut mesomorphic *[sic]* types that appear in commercial airline advertisements. Waino wore thick glasses, was less than 5'10."

Marvin Mangus regularly praised Sig Wein, referring to him as one of the best and most cautious of the Alaskan bush pilots. When the snow began to take flight, Sig, with his plane still on skis, would work his magic. He'd get the most out of the remaining white stuff by using the tundra between the snow patches for landing and take off. And not just any old tundra would do. He chose cautiously and wisely. It always worked; he lived a long life and didn't die in an airplane.

The results of the nine-year program were the proverbial good news–bad news, technical success–economic failure story. The Navy program found recoverable oil in the Simpson oil seepage area of 110 barrels of oil per day (bopd), at a depth of 500 feet. The management team had twenty-six wells drilled in the area, the deepest of which was 7,002 feet. The geology was very complicated, so there was never any reliable correlation between wells, in spite of the fact that the wells were so close to one another.

Several wells in the Umiat area produced oil (Figure 2). The best one produced 200 to 250 bopd for forty-five days. Wells like this in the South 48 would be a bonanza for most operators. Much of the oil was in or near the permafrost, which caused drilling and production problems. Those problems almost certainly could be overcome with engineering studies and trial and error, but the rates of production would not have justified development. For reference sake, the best wells at Prudhoe Bay produced 25,000 barrels of oil per day, and there were hundreds of them.

In retrospect, the well that may have been the most interesting and closest to significant future production was the Fish Creek well, which is only twenty-one miles from production in the Alpine oil field that was developed in the direction of the Fish Creek test, as the Prudhoe Bay production expanded. This Fish Creek well produced only ten bopd from a zone that in all likelihood is correlative with currently producing wells from oil reservoirs of Cretaceous geologic age. It was drilled to 7,020 feet but did not reach the older Jurassic reservoirs producing in the Alpine field. It is problematic that, even if it had reached the Jurassic zones and found oil, it would have been perceived as enough to justify development by the Navy.

Several gas accumulations were found on the North Slope of Alaska, but even today with trillions of cubic feet in the Prudhoe Bay field area, there has not been an agreement on how to get it to market. There is an exception, insignificant economically, except for the village of Barrow. One of the gas reserves was found close enough for the village to use, which made a difference in Eskimo life there. The Navy used that for fuel, which saved thousands of dollars, and in 2002 the field was still producing 234,000 cubic feet per day.

In all, there had been eighty wells drilled in NPR4, thirty-six looking for oil and forty-four that were drilled for geologic information. The deepest test was 11,872 feet. The final report estimated the program had found reserves of seventy-six million barrels of oil. This seems questionable, but even if a certainty, not enough to develop.

Those ten years would present a fascinating story for those who like action, if it had all been filmed. Huge and thick floes of sea ice completely surround a large cargo ship until the appropriate winds free it; three men of a geological field party skidding a heavily loaded boat through the tundra with ropes hitched to their backpack boards; then, those same men run rapids full of terrifying high, standing waves. Tractor trains pull twelve sleds, three with living and eating units, six with drums of fuel and three with miscellaneous freight. Hundreds of tons of equipment and supplies are moved from ships into storage. Hands and machines dig the deep-freeze cavern in the permafrost for frozen meat storage. Seismic shot holes unloading geysers of water when detonated. Clanging of iron and screech of big diesel engines when drilling crew changed a drilling bit. Fire in a moving train of a seismic crew camp extinguished by a visiting supervisor, while the operator driving the tractor pulling the train was oblivious to the incident. A sled-mounted building is made amphibious by placing it on an eighty-four-steel drum flotation raft. A fire destroys the South Barrow Test Well No. 2. Everywhere, there was activity, excitement and drama.

The NPR4 program was a success in that it did prove there was oil in the subsurface, but it never did turn up a clue about the possible existence of the mother lode, the source of the Simpson oil seepages, which was the catalyst for everything connected to NPR4. In all, what was found whetted the exploration juices of oil industry geologists for more exploration in the late part of the 1950s, including the vanguard companies and individuals of Alaskan oil exploration — the pioneers who paved the way to success at Prudhoe Bay.

1

USN
2

3

4

5
31

6

1126
7

8
1110

9

10

11

12

13

14

15

16

17

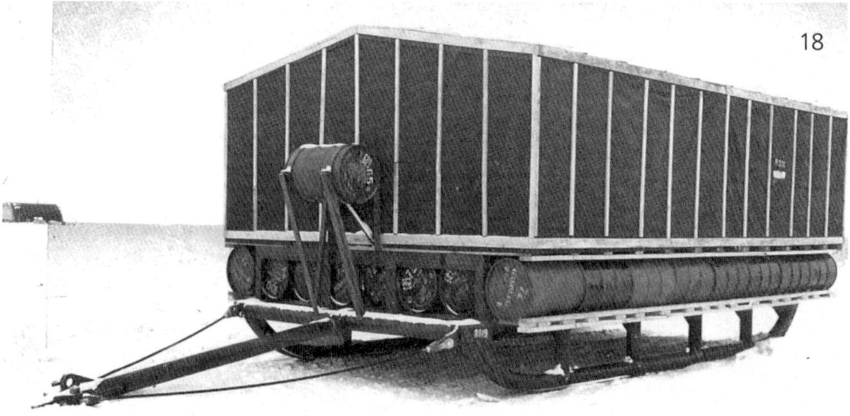

18

19

20

21

22

23

24

25

26

27

28

29

30

31

32

33

34

35

36

37

Chapter Five Photo Captions

1. Each year from 1944 until 1953 the Navy sealift had a narrow window of time in late August until mid September to deliver the cargo to Barrow. Occasionally some ships were trapped in ice (August 1945 here) but always eventually succeeded in getting free. There were no docks, so all the cargo had to be lightered to shore.
   J.C. Reed PP301 USGS898 Photo by USN.
2. The tracked trailers on which the men are standing were called Athey wagons and were used extensively in much of the tractor train operation. Here they were loaded with cargo, which could be delivered to the final site without extra handling. August 1950. J.C. Reed PP301 USGS899 Photo by USN.
3. In the summer of 1945 (the second year of operation) the Seabees constructed this hanger. J.C. Reed PP301 USGS905 Photo by USN.
4. A large carpenter shop was a convenient place to build a Quonset hut. Quonset huts provided most of the living quarters in the main camp at Barrow, the outlying drilling rig locations, and on sleds for the tractor trains, and seismograph exploration crews. The mainframe work of the Quonsets was steel arches. April 1950.
   J.C. Reed PP301 USGS963 Photo by USN.
5. Using barges was another method of bringing the cargo from the ships to shore. August 1947. J.C. Reed PP301 USGS912 Photo by USN.
6. Time was always of the essence because of the possibility of a sudden shift in sea ice that might make unloading impossible, so getting the cargo on shore was of prime importance. There would be time later to sort it. August 8, 1947.
   J.C. Reed PP301 USGS913 Photo by USN.
7. In later years, rather than transporting hundreds of 55-gallon drums of fuel, the Seabees built bulk storage tanks at Barrow. The bulk fuel was carried in bulk storage in the LST and was transferred to the shore bulk storage by pipeline. August 1950.
   J.C. Reed PP301 USGS957. Photo by USN.
8. The LST (Landing Ship, Tank) from WWII was used in later shipping, presumably for its convenience in transporting this kind of equipment to this kind of a landing site. August 1950. J.C. Reed PP301 USGS956 Photo by USN.
9. "Main Street" in the Barrow camp at twilight during the winter. November 15, 1945.
   J.C. Reed PP301 USGS906 Photo by USN.
10. The Jamesway huts here were outfitted on sleds to be towed along the seismograph trails, and tractor trains to provide shelter for the crews.
    J.C. Reed PP301 USGS935 Photo by USN.
11. The men standing in a deep-freeze meat cellar excavated in the permafrost at Pt. Barrow. 1949. J.C. Reed PP301 USGS878 Photo by USN.
12. The U.S. Navy, with their Seabee construction battalions, had been very busy and apparently efficient from the first approval of the NPR4 exploration effort obtained in mid 1944, until the time of this photo on September 11, 1945. Note the large buildings right of center, probably the hanger and other workshops. Talk about a "can do" effort. That was a hallmark of WWII. J.C. Reed PP301 USGS871 Photo by USN.
13. A different view of the Barrow complex on August 8, 1948.
    J.C. Reed PP301 USGS924 Photo by USN.

14. Tens of thousands of drums were used for gasoline and diesel fuel on the North Slope of Alaska. Here are just a few being loaded in sleds for delivery to the drilling rigs and the sub base at Umiat. April 30, 1950. J.C. Reed PP301 USGS948 Photo by USN.
15. This is another type of structure used by the Navy, the Jamesway hut that was similar to the Quonset in appearance, but was more of a temporary structure than the Quonsets. J.C. Reed PP301 USGS908 Photo by USN.
16. The large horizontal piece of equipment is a king-sized pipe wrench, or "tongs" in oil field talk, which is used in conjunction with another one hanging to the right to tighten the drilling pipe seen in the circular area underneath the hands of the man leaning over. This is physically strenuous work and dangerous until one learns the ropes. J.C. Reed PP301 USGS940 Photo by USN.
17. There were several accidents and this is the result of one at a gas well at the South Barrow Gas Field as it appeared on April 6, 1950.
    J.C. Reed PP301 USGS967 Photo by USN.
18. This sled-mounted "wanagan" was another type of building, but frame rather than the arches of steel in the Quonsets. This was a food storage building, which was fitted with sixty-four empty fuel drums for flotation to make it amphibious.
    J.C. Reed PP301 USGS950 Photo by USN.
19. The DC-3 (C-46) with wheel skis did much cargo transporting during all the NPR4 exploration activity. Later, as here on May 1, 1950, the G-2 helicopter was put in service. It had limited capability because of its gasoline-powered engine.
    J.C. Reed PP301 USGS946 Photo by USN.
20. The sleds were loaded eight barrels in length and ten high for eighty per load. At fifty-five gallons each this was 4,400 gallons per sled, and six sleds per train was 26,400 gallons. This train also has two Quonsets for quarters and supplies, and it appears a couple other sleds with other supplies. Please note, the extra crawler tractors. They, no doubt, were spares in the events for travel or mechanical problems. 1950.
    J.C. Reed PP301 USGS947 Photo by USN.
21. This is the massive steel substructure for a large drilling rig. Note the men on top of it for scale. It appears to be resting on piling that will keep the heat generated by the drilling operation from thawing the tundra below. April 28, 1951.
    J.C. Reed PP301 USGS904 Photo by USN.
22. This rig was drilling the Fish Creek geologic structure and was one of the wells that produced oil on a short but sustained basis. It is also the oil that was found closest to what later became the site of the Prudhoe Bay discovery. This was in the summer of 1949, so note the dark vehicle tracks where the tundra has been disturbed and the permafrost below it is thawing. J.C. Reed PP301 USGS880 Photo by USN.
23. This was the location of the Meade River test well and the time was toward the end of winter on April 17, 1950. The frozen ground, compared to that in photo 21, makes for a cleaner looking operation. J.C. Reed PP301 USGS883 Photo by USN.
24. This is the camp and supply yard for the Mead River well. This was another summer operation in 1950. J.C. Reed PP301 USGS970 Photo by USN.
25. A tractor moving a large building with the drill rig mud pumps from a South Barrow well back to the Barrow base. April 19, 1950.
    J.C. Reed PP301 USGS966 Photo by USN.

26. The rope partially seen by the man's left elbow, and twisted around the drill pipe, is pulled by an unseen pulley on a motor and called a "cat head" used to spin the pipe into the joint below it before tightening it with the tongs seen in figure 30. J.C. Reed PP301 USGS926 Photo by USN.
27. This is one of the smaller drilling rigs, a cable tool rig, near the Simpson oil seepages. September 7, 1949. J.C. Reed PP301 USGS941 Photo by USN.
28. This was the location of one of the most westerly of the drilling tests during NPR4 exploration, and was the camp and supply yard for the Kaolak test well. April 28, 1951. J.C. Reed PP301 USGS971 Photo by USN.
29. A heavily loaded "Micheler" sled. 1949. J.C. Reed PP301 USGS928 Photo by USN.
30. In emergencies sometimes the Constellation or DC 4 or 6 was pressed into service. J.C. Reed PP301 USGS929 Photo by USN.
31. Umiat with Colville River in the background. Photo George Gryc.
32. Plane tabling at Chandler Lake in 1940. Walton Banks. Courtesy Barbara Burch.
33. While all the activity of building, moving and drilling was going on, the USGS geologists were conducting geologic exploration, in this instance, dragging a boat from one area to another in the course of their work. J.C. Reed PP301 USGS872 Photo by USN.
34. Supply tent with Eskimo in attendance. Walton Banks. Courtesy Barbara Burch.
35. Preparing to leave Chandler Lake to begin surveying toward the Colville River and the Arctic Ocean. Walton Banks. Courtesy Barbara Burch.
36. The other is a Gryc field party camp on a river bar, 1945. Photo George Gryc.
37. Piper Cub supply plane landing on a river bar, 1945. Photo George Gryc.

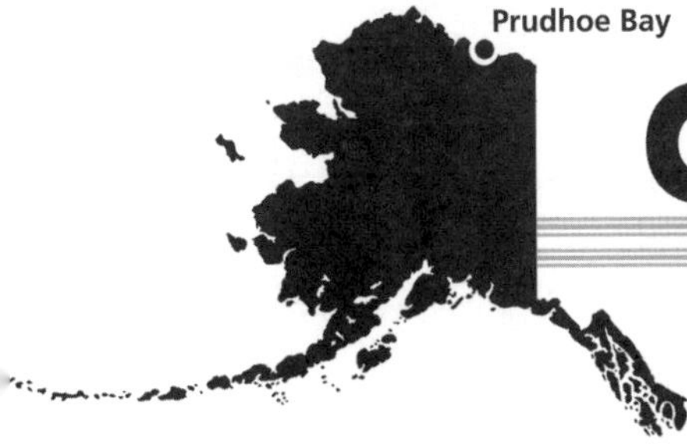

Chapter 6

Richfield Discovers Commercial Oil — 1957

Geologists have an overpowering love affair with the unknowns in the guts of Mother Earth. An insatiable spirit is kindled and sparked to flames of passion when optimism and enthusiasm fuel it. The following humorous story makes the point.

A geologist was at St. Peter's gate, hoping for entry. St. Peter suddenly had to stand aside as a rush of people was leaving Heaven. The geologist, looking bewildered, asked why they would want to leave. St. Peter said they were geologists going to hell. They had heard a rumor that there might be oil down there. The waiting geologist turned to join those leaving. St. Peter reminded him, "It was just a rumor." The geologist replied, "So, it might be true."

In his book, *Crude Dreams,* Jack Roderick gives a good history of oil exploration in Alaska from day one. For many years there had been a remarkable number of different exploratory efforts with only fleeting economic success. In the 1950s the oil industry began to get renewed interest in Alaska. Until 1957, some senior managers probably thought their geologists were going to hell for wanting to explore in Alaska. That all changed when Richfield discovered the Swanson River Oil Field (Figure 10) on the Kenai Peninsula in 1957. Swanson River oil field was still producing a modest amount as recently as 2004. It was a major discovery and has produced more than 225 million barrels of oil. "Major" usually means anything over 100 million barrels. Prudhoe Bay in 1957 was not even in any oil geologist's vocabulary. It is roughly 800 miles almost due north of Kenai but, little by little, with piece by piece of exploration evidence, industry geologists found their way to Prudhoe Bay. The exploration spirit led them there. A renowned petroleum geologist once said, "Oil is found in the minds of men." Men and women now have to put all the pieces together to make discoveries happen.

Without Richfield, oil may never have been discovered on the Kenai Peninsula. On the other hand, the seismographic evidence of the future oil fields in the waters of Cook Inlet was so strong that

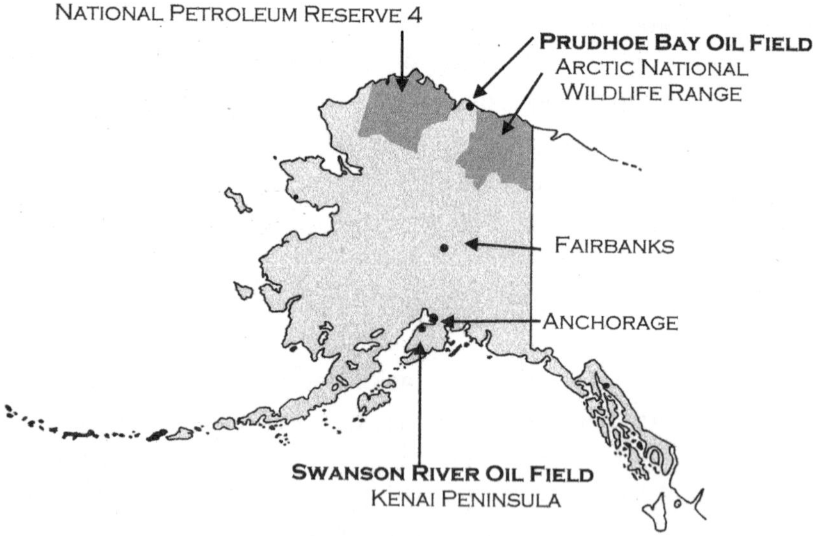

Figure 10: Swanson River oil field begets Prudhoe Bay oil field. This shows the geographical relationship of the Prudhoe Bay oil field on the Arctic coast, to the Swanson River oil field on the Kenai Peninsula, southwest of Anchorage. After Specht, R.N., et al. Geophysics, *Vol. 51, No. 5 (May 1986) SEG (Copyright) 1986 reprinted by permission of the SEG whose permission is required for further use.*

one of those fields almost certainly would have been discovered. Some other company might then have the honor of opening Alaska to intense oil exploration. Richfield's discovery at Swanson River, and Richfield's purchase (with Humble Oil and Refining Co.) of the leases under which Prudhoe Bay oil field (Figure 10) was discovered, led the vanguard in establishing Alaska's major contribution to the national energy requirements. As an aside, and indicative of the vagaries of oil exploration, there has never been any other oil discovered on the Kenai Peninsula. It has been explored thoroughly with seismic and drilled many times. This is not to imply that more oil won't ever be discovered, it just seems unlikely.

The Swanson River oil discovery, besides being major in a quantity sense, was equally important because it finally proved that there was economically producible oil — at least in the Cook Inlet basin. William C. Bishop, a Richfield geologist, is memorialized in the museum in Anchorage by the bronzed boots he wore when selecting the location of the discovery well for the Swanson River oil field. The location was based on only one seismograph survey line. The likelihood of one survey line in an unexplored area giving an indication of where to drill is virtually nil. Ideally, a grid of lines is necessary to show the big picture three dimensionally. Frank Tolman, a Richfield geologist, had observed a large topographically elevated land area in the northern part of the Kenai Peninsula. Historically, topographic highs and river bends have on occasion been caused by subsurface geological features that are conducive to oil accumulations. The odds for success based on these two weak bits of information were nebulous. The discovery well found the proverbial needle in a haystack. The field was large, and eventually covered several sections, but the discovery had been drilled in what turned out to be the northern tip of the oil field. Had Bishop dug his heel in the ground 2,500 feet north, east or west of where he did, Richfield would have drilled a duster.

The real justification for drilling the wildcat test was more a business decision than one based on strong geologic evidence. The odds against a discovery were so great, one might wonder if it was a predestined event or divinely led.

The book *Crude Dreams* describes in detail what took place, but in summary, Richfield began to acquire federal oil and gas leases in the north Kenai area on the basis of Frank Tolman's observation. Locke Jacobs, an enterprising young independent oil and gas lease broker, noticed the Richfield activity in the public records and began to make his own lease applications in the same general area. Jacobs wrote to Richfield's land department in Los Angeles, asking if Richfield would be interested in discussing the area where Richfield and Jacobs leases intermingled. Jacobs represented a group of Anchorage businessmen. What happened was the result of pragmatists in both the Jacobs and Richfield camps wanting to make things happen that were in their respective best interests. The backers of Jacobs were really interested in generating business activity, not in making money directly from the possible sale of oil leases. It was, and still is, common for oil companies to use a strategy of drilling a well in return for earning an interest in leases. Richfield wanted to earn an interest in hundreds, more likely thousands, of acres by drilling an exploratory well. That is what Richfield offered, and the Jacobs group accepted. In addition to drilling, Richfield did pay the Jacobs group a small bonus for each acre they leased and an overriding royalty of 5 percent in some of the leases. An overriding royalty is an additional royalty negotiated on a lease when it is perceived to have extra value. These were U.S. government leases, and federal lease stipulates the leaseholder will pay the U.S. government a royalty of one-eighth or 12.5 percent. Richfield was willing to pay to the Jacobs group another, overriding royalty of 5 percent, over and above the customary federal royalty. Twenty-five years later, this 5 percent override paid the seven or so members of the Jacobs group an estimated $3 million each, and the money continues to flow as long as the oil does.

Only weeks after the discovery, Chevron paid Richfield $30 million for a half interest in all the Richfield leases on the Kenai Peninsula, including the leases in the discovery area. What a marvelous business decision Richfield had made. At this point no one knew the potential size of the oil field, but Richfield had $30 million in the bank and a half interest in whatever was to come.

During the drilling of the discovery well, G. Ray Arnett was the well-site geologist. In 2002, when asked what he remembered about some of those events and any interesting or colorful incidents, he remarked, "The best color remembered from those heydays of 1957 are not fit for a family publication." That lone comment speaks volumes about Ray Arnett.

Here are some things he remembered that are fit for this story. It will give readers some sense of what is involved in drilling an oil exploratory well — a "wildcat" in oil field vernacular.

> Drilling the *Swanson River Unit* No.1 wildcat resulted from a number of factors. Bill Bishop, Mel Sweeny, Ben Ryan, and I, all Richfield geologists, had conducted summer field mapping in the area in 1955. In addition to field mapping, a geophysical program was carried out very carefully within what was then known as the Kenai Moose Range, so as to minimize to the greatest extent possible, any ecological or moose habitat degradation. Although geophysical and field mapping results were poor, they did provide "indication" of possible structure. Stronger indication came from [other] geophysical evidence and aerial photographs that show the Swanson River's flow toward the Cook Inlet was diverted, either by glacial moraines blocking the normal flow, or subsurface structure. Cal Tech glaciologists carefully studied the photos and could come to no definite conclusion as to why the Swanson River course was altered eons ago.

Richfield was an aggressive, great little company, and its exploration budget was not unlimited. As I recall, Frank Morgan, Richfield's VP of Exploration, had his eye on prospects in Australia, but [by diverting funds] he went along with exploring in Alaska, so that is how the funds became available to drill the SRU#1 wildcat *[Swanson River Unit No.1,* is the discovery well's official name].

The 22-mile oilfield road used to get the drilling crew from Sterling highway turnoff to the rig location was built during the fall and winter of 1956 when temperatures fell to 70 degrees below zero. With the coming of spring in 1957, in order to beat the Alaska road ban we hustled unloading the barge that brought the oil rig parts and equipment from California to Seward. Many loads had to be trucked to the well-site location and unloaded. Loads of heavy steel rig parts, drill pipe, drill collars, drill bits, thousands of sacks of drilling mud, heavy cables, the rig's large diesel engines, fuel storage tanks, sheets of galvanized tin roofing for the core shed, a large live-in trailer the tool pusher and I used for sleeping quarters, etc. Beating the Alaska road ban was one thing, but our 22 miles of new gravel road was thawing too, which compounded the difficulty of moving in and assembling the rig and location.

The crews worked two tours [pronounced "towers" by oil field people] of 12 hours on and 12 off. I once asked a roughneck why "tour" was pronounced, "tower." He said, "How the hell do you pronounce "sour"?" That was good enough for me. Anyhow, two 12-hour tours were better than three 8-hour tours because the crews had to be shuttled back and forth from Soldotna where they were quartered. An Army surplus 4 X 4 was used to shuttle the crew. The truck was fitted with large oversize airplane tires to enable the truck driver to negotiate 22 miles of oilfield road we had bulldozed through the wilderness of tundra and spruce trees. It was springtime and the road kept falling apart as the frozen road gravel and sub base soil thawed. The road required constant maintenance and repair.

Geologist Arnett was responsible for examining and describing the well cuttings as they were brought to the surface by the drilling fluid ("mud"), keeping a daily log of drilling progress, and reporting daily to the home office in Los Angeles. He would identify and describe the character of the rock. The description would mention, its texture, color, grain or crystal size, whether it was sandstone, shale, limestone, coal or basement rock (crystalline rock with no oil potential), the minerals present in the cuttings or core samples, the mineral grain size, the porosity and permeability of cores, the color and most important of all, watch for evidence of oil or gas. On a remote wildcat well like *SRU* No.1, it is not uncommon for the well-site geologist to share in some drilling decisions with the drilling engineer. In this regard, Arnett wrote an interesting letter to Bill Bishop in the Los Angeles office, detailing the problems and gyrations he had to go through to get drilling mud delivered to the discovery well site after drilling was underway.

The drilling crew on the *SRU* No.1 saw a problem develop as the mud became diluted by gas coming from numerous thin layers of coal beds in the drill hole. There was a need for many tons of 100-pound sacks of weight material to recondition the drilling fluid (mud) to counter the gas entry into the hole. Drilling mud is pumped down the drill pipe to cool the turning drill bit, and then circulated back up, carrying the rock cuttings from the bottom of the hole to the surface. The mud flows through the screen to continue circulating, but the cuttings are caught on the screen and are examined by the geologist, then saved. During the drilling operation, the mud continues its endless journey as it is pumped in and out of the hole. A mud technician also continually monitors the mud with on-site lab instruments to control its character to return cuttings to the surface and maintain a weight to prevent a blowout from high-pressure gas or lower pressure coal gas.

The mud needed Baroid, weight material to control formation pressures. There were no oil field supply stores in Alaska, but Arnett was able to negotiate for the needed weight material from a Phillips Petroleum wildcat operation that was being abandoned in southeast Alaska. Finding it was the easy part. Transportation being what it was, he also had to arrange for the material to be barged from the Phillips Petroleum location to the Kenai area, a distance of about 400 miles. Here are some of Arnett's travails in making arrangements for getting the mud delivered.

> About 5:30 a.m. Sunday, the 19th, Phillips called and said we could have the material. That's when I got started trying to locate a barge to carry it for us.
>
> I'm sure you can appreciate the trouble we had. The fishing season was just starting in Cordova [near Katalla where the material had to be loaded] and none of the fishermen wanted to release their boats to haul the Baroid [the brand name of the mud]. The earliest one [barge] was expected back in Cordova was the 22nd. Well g - - d - - - that wouldn't do because that meant five to seven days before we would have the Baroid at the rig. ...Bob Logan, marine insurance adjuster, etc., from Cordova was in Anchorage...Logan flew to Cordova and scrounged around for us. He called...and said he thought he had a small [30-ton] barge that may haul the stuff for us. By 5 he called back and said that one barge Captain had crapped out on him but he had gotten another one who would haul it, strictly as favor, for $1000. I had to dead head two Weaver Brother trucks from Anchorage to Seward to haul the load to the [drilling] rig.

Everything was going fine until the barge arrived in Seward about 6 p.m. on the 20th. Then the offal hit the blower! What with all the [expletive] stevedore union regulations, Alaska railroad charges and stipulations, the unwillingness of the cold storage to handle that type of cargo because of their fear of the unions, etc., nothing could be done until 8 am on the 21st. So that meant the rig, trucks, barge, and all of us sat on our hands another 14 hours [all chargeable at high Alaska rates]. You see, the barge only had 620 sacks and Jones figured he would need at least 1000 or 1200....to save from 3–5 days of rig time, and the possibility of losing the 9225 feet of good hole we already had. I finally got a firm commitment from Cordova Airways to fly two loads. Then Jones called from Seward to say they couldn't get started unloading until 1 pm because railroad cars were on the way. After that the longshoremen refused to use the railroad crane on the dock so we had to contract another crane. What a lot of BS! It still appeared we were on schedule. But I'm a SOB if Cordova Air Lines didn't call and give me the great news that both their C-46s [planes] were unavailable. I was fit to be tied. Honest to God, I was at my wits end! To shorten the story Logan got another barge and better docking and unloading arrangements in Seward for the 2000 sacks of baroid (100 tons) delivered for $1300. The first shipment of 620 sacks had cost $800 plus charges for moving the rail cars.

ROCO [Richfield Oil Co.] has some very fine friends here and we should feel fortunate. Without this kind of congeniality and unselfish assistance it [would] cost a lot of money and time to try to get by up here.

Mr. Arnett describes other problems during the drilling of the Swanson River discovery, documenting the consequences of operating in an area remote from the rest of the industry in the Lower 48. Once, Arnett wanted to take down-hole measurements that required lowering instruments down the hole on an electrical cable. In addition to Arnett, there was Eston Jones, the Richfield tool pusher. The tool pushers are in charge of the rig, the crew and the well drilling operation. They are responsible for drilling a good, safe hole, so they are not always amenable to the desire of the geologist to evaluate the evidence of oil and gas "shows" in the cuttings, the drilling mud or a core.

The down-hole formation testing just mentioned was scheduled to be run by Schlumberger, an oil service company contracted from California to do the work. After consultation with Arnett and the Schlumberger crew, Jones decided the drilling mud wasn't in good enough condition to chance running the electrical test. Because no test would be run, Arnett released the Schlumberger crew but told them to leave the testing equipment on site. The two-man test crew jumped into their car and started the long, rough trip back to Anchorage.

Arnett vented his frustration.

> About 10 a.m. that same morning Jones [the drilling engineer] came in from the rig and said Bakersfield [California and the supervisors there] had called and asked him if he thought it would be OK to run a log [the instrument on the cable]. Jones told them he though it would be! Hell, I'd just sent the men away because the night before he thought it wouldn't be OK.
>
> Well anyway, I sent Rakestraw [a Richfield petroleum engineer] in my car to try to catch them. I called the Territorial Police to try to stop them on the Seward Highway, and I got Harry Reger to fly me in his plane to try and head them off. I had a note wrapped in large rag with some rocks, to drop to them. They had about an hour's start on us and Rakestraw and

the police didn't catch them and Harry was running into bad weather and out of gas going in the pass so I told him it was OK to turn around and come home.

The Schlumberger crew made it to Anchorage and, unfortunately for them, when checking into their hotel for a well-earned good night rest, they were handed my message to return to the rig, post haste. That sure as hell wasn't welcome news for them. It was a long, hard car trip from Anchorage to the rig site in those days. There were few paved roads into or out of Anchorage, and the Sterling highway was more like a wagon trail with miles and miles of rutted gravel surface all the way. I can't tell you how many mufflers were torn from our cars traveling those Alaska roads.

I've got my fingers crossed we can find something on the EL [Electric Log, the instrument on the cable] and still have something to test. We've had a hell of a lot of gas shows since 7940 [feet of depth]. True its methane and maybe not commercial, but if we're not optimistic we don't belong up here.

Forty-five years later Arnett added the following.

What would have been a relatively simple operating procedure and easily accomplished under normal oilfield conditions at Lost Hills [in California] or Pecos, Texas, was a monumental task on the Kenai Peninsula in 1957.

All well reports and other messages had to be transmitted by radio to Anchorage, then by radio to Juneau where the transmission continued via phone lines to our Los Angeles office. Before coming to Alaska, a secret code had been developed for sending drilling reports, because reports sent over radio could be picked up by oil scouts and others who might be interested in what we were finding. Richfield had a big stake in the *SRU#*1 wildcat and it didn't serve her purpose to allow others to know where we were in the hole, what formations had been encountered, how the drilling was proceeding, etc.

Our well-site radio's antenna was mounted above the rig's crown block at the very top of the drilling rig. The antenna should have been mounted on a stand-alone tower, but to save a few bucks the rig's derrick served as the tower for our antenna location. In retrospect, this was penny wise and pound-foolish because the vibrations in the derrick when drilling continually caused the antenna arms to break loose. Just about all of the 3-foot long pieces fell to the derrick floor like aluminum arrows, endangering the roughnecks [the drilling crew].

I called Anchorage for a radioman to be sent to the rig. His job was to secure the antenna and its cluster of radiating arms to the top of the derrick to eliminate the problem of falling "arrows." The problem was explained to him. He walked back to his truck for needed tools, and I went about my usual morning routine. Several hours later, I noticed that the radio technician's car was still parked near my trailer, and he had not reported back to me. I started to wonder about the guy. The driller reported that he'd not seen the radioman since the guy had arrived. A couple more hours and I became more concerned. The radioman's car hadn't been moved and none of the roughnecks had seen him. I decided to climb the rig ladder to the crown block and see if the antenna was fixed. As I started up the ladder, about two-thirds the way to the top there was the radioman, paralyzed with fear, his grip locked onto the ladder rungs, and hanging on for dear life. If he had ever called for help, no one heard him. When drilling, loud clanging rig and engine noises would have drowned his cry for help anyway. The guy was just hanging there — like a date on a calendar — scared spitless, glued to the derrick ladder, and about 80 feet off the ground.

Since the guy couldn't be talked into letting go, I climbed up to him to see what I could do to help him down. The man was speechless and I was unable to pry his grip loose while hanging onto the ladder with one hand. I called for the derrick man [he works in the derrick when the drilling pipe is run in and out of the hole] to come up and climb over us so he could pry the technician's fingers and hands loose from above. I stayed below the man in order to move his feet, one at a time, off one rung and then put it down on the next lower rung. One hand and one foot at a time, the three of us inched our way down the ladder until the guy could regain his confidence and was able to help us help him down to the derrick floor. Needless to say, the guy was embarrassed, but grateful to be on firm footing again. The derrickman, under the direction of the radioman, finished the job of securing the antenna.

The foregoing difficulties not withstanding, the discovery well, Richfield Oil Corp. *Swanson River Unit* No.1 was spudded (had begun drilling) on April 5, 1957, had a formation test which yielded oil on July 19, and was completed August 30, 1957. That was good time considering the remoteness of not only Alaska, but particularly the well location within Alaska. Up until this discovery there was no oil field infrastructure anywhere in Alaska. Everything drillers might reasonably need would have to be brought with them, or they'd suffer the high expense of delay and transportation. Drilling contractors always protect themselves by requiring payment during times when, through no fault of theirs, the rig remains idle. The operators of each drilling venture had to anticipate what might befall them during drilling and to be prepared for those eventualities. There would have been no way to prepare for every contingency without breaking the bank, so Richfield took calculated risks that the unexpected would not happen. Richfield was fortunate when they did encounter the unexpected occurrence of having to find and ship the Baroid drilling mud sacks over 400 miles. In the grand scheme of things the cost was not excessive, even back then.

Here is how Arnett remembers what occurred, step by step, until he and the crew saw oil.

> On the way down [as the well was drilled from the surface deeper and deeper] to the pay zone, [where the oil is] there were many gas shows [indications of gas in the drilling fluid obtained from detection instruments] caused by drilling through thin layers of coal, but who knew? We always stopped drilling to circulate out cuttings [the drilling fluid carried rock chips to the surface] when a hydrocarbon kick [indication] showed in Core Lab's wagon full of equipment. [This enabled Arnett and the Core Lab crew to evaluate the rocks in which they were drilling for oil and gas indications from the mud, and cuttings from the rocks being drilled.] When the pay zone was reached the [rock] cuttings had traces of oil, so I had the driller run in [go in the hole with equipment] for a core. The driller replaced the drilling bit with a diamond-coring bit, which cuts a sample of rock. This usually yields pieces of rock four inches in diameter and from a few inches to several feet long, which then is examined and described by the geologist. If warranted, the zone from which the core came may be subject to further testing for oil and gas. This [the time required to stop drilling and start coring] gave

me an opportunity to wrestle the 22-miles of gravel road then the main road to Kenai to get a decent dinner at 4 Royal Parker's bar and restaurant.

As I recall, it may have been around midnight when I returned to the rig. The core barrel was just coming out of the hole and being laid down on the catwalk [a long ramp from the drilling rig floor used for drill pipe storage] so the core could be examined. Before the core was dumped out, I looked into the end of the barrel [coring tube], scraped out a bit of the core and was elated to see it contained oil. I don't recall how many feet of oil sand were in that core. Not much, but enough to run a JFT. [A Johnson Formation Test is a drill stem test of the formation from where the core was cut. This is done with a system of strong rubber sleeves, called packers, added to the drill collars and drill pipe. Once on bottom, weight from the drill pipe expands the packers against the drilled hole's sidewalls to isolate the drilling fluid outside the drill pipe and above the packers from the zone being tested. Systems of valves then are opened. If there is oil and gas present, those taking the test hope the zone pressure is sufficient to bring the hydrocarbons to the surface.]

Once on bottom and the packer set above the test zone, the bar was dropped [this opened the test equipment at the zone being tested] and we waited, and we waited, and we waited. Eston Jones, the Richfield tool pusher responsible for the rig and drilling operation, was getting antsy about the packers, drill collars, and drill pipe getting stuck in the hole when the packers are set too long. When pipe gets stuck in a hole, an expensive fishing job [a complex procedure of pulling (fishing) the down-hole packers and drill stems out with the hope of saving equipment and the hole] is required to try and recover the stuck pipe. Jonesy wanted me to give the word to pull the packers and come out of the hole.

Although we were getting only a very light blow at the surface [this indicates something, oil, gas, or formation water, is entering the drill pipe, thus displacing air from inside the drill pipe, which is causing the blow] I'll admit that the blow was very faint. I had placed a wet, muddy rag over the drill stem to detect even the slightest puff. I said, "Patience, Jonesy. Fluid is entering the pipe. Pretty weak, for sure, but let's give it another hour or so to surface."

We'd had so many false shows drilling through coal beds on the way down to the oil zone that Jones said, "S- - -t! We probably just drilled through another dead moose and that's where your $@#% gas is coming from." He was worried, and rightly so, about sticking the packers, which would entail an expensive "fishing" job ahead. If unsuccessful, that might end up losing the hole and leaving everything [failing to recover the stuck equipment] below the packers in the hole. If a strong enough pull is applied to the drill string, even the drill pipe itself can be pulled apart. When a fishing job is unsuccessful, the hole is cemented off. [This lower part of the drill hole would have to be abandoned and, if worth it, efforts may be made to re-drill the abandoned part of the hole.] Tool pushers don't like to think of that. [Geologists don't either, because it is costly, uncertain, and interrupts the flow of unfolding unknown geologic information. This situation occurred later during a critical juncture when drilling the ARCO *Prudhoe Bay State* No.1].

For the next several hours it was a cat-and-mouse game between Arnett wanting to keep the test going and Jones wanting to stop.

This was a common occurrence in the oil fields. Here are parts of what went on from Arnett's perspective; one can be sure Jones's perspective was different.

> When I saw him [Jones] coming my way, I'd scoot around the rig site in another direction. It got so I was hiding among yard equipment, behind trees, and in the core house to avoid Jones. Nevertheless, I could hear him yelling. "Arnett! Hey, Arnett! Where the hell are you? We've been on bottom too g- -d - - - long already. Let's pull the $@#&% packer and stop this bull s - - -. You hear me, Arnett? Where the hell are you?"
>
> This charade went on a long time, but each time I was near the rig floor I would check on the wet rag over the drillstem while Jones was in the trailer, or the doghouse, or outside not finding me.
>
> After a long period of waiting, there eventually was a stronger indication that fluid was indeed nearing the surface. I wasn't about to pull the packer loose until I saw what the drill pipe contained. Finally, oily water surfaced into the mud pit, followed by a weak flow of oil coming in spurts as the fluid began to clean up [become all oil]..
>
> We had already bulldozed a holding tank nearby in the event we were fortunate to find an oil zone. In time, the flow cleaned up and just crude oil flowing heavily. Then is when the flow was diverted from the mud pit to the crudely dug holding tank.
>
> I had strapped the pit [measured the height, width, and depth] — and made several wood stakes with inch marks penciled on them to stand up in the bottom of the tank. This allowed me to estimate the oil flow as the tank filled and the fluid level inched its way up the stakes over a given period of time. My estimate

> was between 800–1050 barrels/day. As I recall, the official rate, calculated with sophisticated oilfield equipment, was 900 b/d. [The official completion flow rate was 900 barrels of oil per day after permanent well casing (pipe) and the Christmas tree (a series of control valves at the wellhead) were installed, and *SRU*#1 was on production]. My calculation was not bad, considering I only had a bulldozed hole in the ground and some wooden stakes penciled with 1-inch marks. In a normal drilling operation, a strapped (precisely measured) holding tank would have been available. [A strapped holding tank would have revealed a precise measurement of the oil flow.]

Near the end of the test when the flow had increased greatly, what would have been a tragic accident could have occurred. Gas was collecting like a fog in the entire five-acre cleared area surrounding the well site. A small spark could have turned the area into an inferno.

> Then I saw a flash go off! One of the crewmembers had taken a flash picture of the oil flowing into the pit! Jaysus! An explosion and a firestorm was all we needed to put an end to an otherwise grand occasion. I never knew, and didn't want to know, who took that flash, but it was the last flash picture taken that morning.

The immediate effect of this discovery caused a flurry of exploration and drilling the following few years, which resulted in many successful drilling ventures in the Cook Inlet. In keeping with its early success, Richfield continued to participate in other oil discoveries in the Cook Inlet. Several companies chose to operate in partnership in Cook Inlet. Richfield became a member of the Shell–Richfield–Standard of California group. This group was very active and had great successes at Beluga River, Middle Ground Shoal, North Cook Inlet and Redoubt Shoal fields.

The optimism for oil and gas discoveries on the Arctic Coastal Plain began to awaken almost immediately. The American Association of Petroleum Geologists each year publishes in its bulletin the important exploratory drilling events of the preceding year. The Alaska review for 1958, reported that awakening with the first oil industry geological field party activity in Arctic Alaska. Richfield heeded its wake up call and sent its first geological field parties there in the summer of 1959.

Richfield continued its interest on the Arctic slope in 1963 by sending Gil Mull and Gar Pessel to the Umiat–Chandler Lake area to expand on the work done by the USGS for the Navy on NPR4. They were accorded great trust and confidence to pursue their work as they saw fit by their supervisors including, Mason Hill, Manager of Exploration and his assistant, H.C. Jamison in Los Angeles. Mull and Pessel spent virtually their entire careers working the geology of Alaska. This was the beginning of the association of H.C. "Harry" Jamison with the unfolding of the Prudhoe story, continuing through the discovery, confirmation and construction of the Trans-Alaska Pipeline.

Mull and Pessel used a Bell G-2 helicopter, and a fixed-wing Cessna 180 for transportation. Helicopters were light-years better than the fixed-wing air support, weasels, canvas boats and shoe leather the USGS had to rely upon during the NPR4 exploration. Helicopters obviated wading in ice water to the armpits, but were no less dangerous. Carl Brady, a pioneer helicopter pilot and owner of a helicopter service company, told me a fixed wing (airplane) is safer than a rotary wing (helicopter) because a landing field is stable, and much larger than most of the small landing targets of a helicopter. Obviously there isn't room for an airstrip on a drilling platform, a drilling ship bobbing in the swells, or a small boulder-infested bench on a mountainside. The early piston-powered helicopters were severely limited as to the weight they could carry and the altitudes at which they could fly. Several engine performance parameters had to be within narrow limits simultaneously for safe operation. Those parameters were not always easy to keep within their ranges and occasionally pilots would be in a position of having to compromise, or worse, gamble if there were no other options. For example, if one is hovering in a mountain location, dropping off or picking up a passenger, and the readings aren't right, the pilot doesn't have the option of doing anything but flying out of that situation. Turbine engines, with greater power than piston engines, were a great advance in performance and safety.

Gil Mull never had a close call flying, but once when Harry Jamison, Charles H. Selman and some others visited a seismic crew on the Arctic Slope, the plane that came to take the group back to town was smaller than the one that brought them. Gil Mull related the following.

> In mid-December 1963, several of us, including Harry Jamison, went up to the Slope to spend some time with Richfield's first seismic crew there. So just before Christmas, when we got ready to leave, we had to fly back to Umiat to catch a DC 3 flight back to Fairbanks. A Pilatus Porter — a Swiss-made, high-performance plane on skis — landed on the snow close to the seismic camp to pick us up, but the pilot said that with the load and short strip, he would not have room on board for all of us. There was a smaller plane, a Cessna 185, hauling some freight in to another strip located on the ice on a lake 6 or 8 miles south of the seismic camp, so I volunteered to go down there and ride back to Umiat with that pilot. I hopped in a helicopter and we flew down to this lake to wait for the Cessna 185. We sat there waiting in the helicopter on the lake with the engine running for a few minutes, and finally I got out just to stretch my legs a bit. The next thing I knew, without saying anything, the helicopter pilot just picked up and

left, leaving me in just my parka and cold weather gear standing there on the ice in the middle of that lake in the mid-day twilight at –40° with nothing around but a few fuel drums and absolutely no shelter. The pilot apparently assumed that the other plane was going to be arriving shortly, and he had someone else to move. He hadn't been gone for more than a few minutes, when the wind came up some and a whiteout moved in, so that there was almost no visibility — like standing in the middle of a full milk bottle. This was suddenly a serious situation, because in a whiteout [in this situation there is no horizon or points of reference; one sees only the sameness of white or shades of gray] I had no reasonable expectation that the fixed-wing pilot would be able to locate this solitary lake [in reality, there usually is a plethora of lakes] in the middle of the tundra. I was beginning to consider my options, and had begun to look at the fuel drums to see if by chance any of them had fuel that I could get out and perhaps light for a fire, when out of the gloom, in came the 185. The pilot was shocked to find someone waiting there for him. He had absolutely no idea that he was going to be taking a passenger back to Umiat — no one had called him on the radio to tell him that, and given the whiteout, he could very easily have decided to go back to Umiat without completing his trip.

Neither the people at the seismic camp nor those who went ahead would have had any inkling of Mull's tenuous situation. Had anyone known, darkness would have delayed a rescue attempt for many hours until the next short period of twilight. Gil says he owes his life to Bob Jacobs, the pilot.

In *Prudhoe Bay, Discovery to Recovery* by Gene Rutledge, H.C. Jamison credits Gar Pessel with providing the original impetus for Richfield's big plunge into the Arctic Coast. In a letter written from Gar on yellow lined tablet paper dated August 2, 1963, he wrote,

> The most outstanding feature to date has been the nature of the tertiary formation between the Tulik and Kavik rivers. These sands are not only good but also more correctly called fantastic. The P&P [porosity and permeability] of these sands can be called good to excellent and we are dealing with a section of at least several hundred feet net of such sand.

Gil Mull says they, "found outcrops of sandstone, which correlate with sandstone found several years later in the Kuparuk producing area near the Prudhoe Bay field." The Richfield geologists were now wide-awake and fully energized.

One or more of the many USGS parties had in all probability seen these favorable rocks in the 1940s and early 1950s. Nonetheless, the staff's confidence level was heightened when one of Richfield's own made such reports. Based on this, H.C. Jamison recommended to his supervisor, Mason Hill, Manager of Exploration, a seismic program to obtain subsurface structural information with a cost of just under a million dollars. That doesn't sound like much now, but for any company back then that was big money, especially for Richfield. Mr. Hill agreed, but then left on an extended trip, so in his absence Harry Jamison, now Alaska Exploration Manager, had to continue climbing the ladder for that kind of money. Frank McPhillips, General Manager of Exploration, was the next stop. Mr. Jamison describes him as a visionary and always aggressive. McPhillips had advanced through the land department. Land departments are responsible for buying leases and handling many of the business decisions leading to drilling. He, too, became a proponent.

The next level of necessary approval led to Bill Travers, Vice President of Exploration and Production. Mr. Travers came up through the ranks as a petroleum engineer. Engineers are trained to deal with facts and often find it difficult to deal with the unpredictabilities of exploration. They become uneasy with uncertainty, but Travers was intelligent and forceful, open-minded and willing to listen to those who had demonstrated good thinking and judgment. He was also a risk taker. Successively higher managers have a broader area of responsibility; consequently each project brought to them for approval must compete for funds with more and diverse departments. Mr. Travers had to weigh Mr. Jamison's recommendation against all the exploration and production needs across the entire Richfield Corporation. Those subordinate to Travers had long since established their bona fides with him. Because of this, and the unanimity of the proponents in favor of the expensive seismic program, he gave the final approval. A United Geophysical Party, supervised by Peter Gathings, began surveying in December 1963.

Pete Gathings soon became known in Anchorage as "Mr. United." At one time or another he supervised seismic crews under contract for most of the companies exploring the North Slope. Pete prevented disaster on one of the early crews United had working for Richfield. Here is Harry Jamison's account.

> We were moving [the geophysical] camp. A large Caterpillar was in front pulling a string of wanigans [portable sleeping, bath, kitchen, and dining units mounted on sleds] to a location nearer to where the seismic crew was shooting [surveying]. I remember the drama. A fire broke out in the kitchen and was raging. Pete Gathings saw it and fearlessly grabbed a fire extinguisher and, running alongside, got inside and was able to put out the fire. It was really dangerous because the flames were licking the propane tanks, which were inside the building. The guy driving the tractor had no idea what was happening and so the train just kept sliding across the tundra. If Pete had not succeeded, there was a good likelihood the

> whole camp would have been lost, or at very least out of business without a kitchen.

In this remote and often forbidding and unforgiving place, those involved had to be resourceful and do what needed to be done.

Charles H. Selman came on the scene in 1961 as District Geophysicist with geophysicists Jim Clinton, Peter Clara and Rudy Berlin reporting to him. These three men interpreted the records as they came from the field. At that time the data from the seismic crews in the field were wavy lines on paper three or four feet long and four or six inches wide. From these they would make maps of different rock levels below the surface. All three of them made maps of the Prudhoe Bay structure, based on field records at the time of their interpretation. All of their maps agreed in overall character. Only the details differed.

Charlie Selman stayed with ARCO until retirement. His death in the fall of 1999 provided the motivation to begin this book endeavor, and it is to him and the others who have passed on that it was dedicated. Charlie had had some health problems, but died unexpectedly in his sleep. His death was untimely in the hearts and minds of his contemporaries. He was jovial and a great story teller, and could have contributed greatly to this story. His audible all-purpose trademark explanation was a forceful “Ha” with a short and hard “a.” Charlie’s wish was that his ashes be scattered at the mouth of the Sagavanirktok River near the Prudhoe Bay oil field, and they were.

Frequently, companies will join forces in a frontier exploration area like the Alaska Arctic Coastal Plain was in the early 1960s. The objective is to spread the risk. At this time Richfield was a comparatively small company operating in remote Arctic Alaska in a high-stakes game. Humble Oil and Refining Company was very large and financially strong and was interested in becoming active there. Rather than start from scratch, J.R. Jackson from Humble approached the Richfield counterparts and asked if Richfield would be interested in having a partner to share the costs and any future successes. Richfield liked the idea so they negotiated a joint exploration agreement in 1964. Richfield was named the operator. Humble paid Richfield for seismic work and leases Richfield had acquired. Up to this point Humble apparently had not seen much seismic detail, and this was not unusual at this stage of a newly formed joint operation. According to a Humble source, the Humble exploration people were "dismayed" by the lack of geologic structures when they did see the geophysics. Presumably this was because Richfield may not yet have mapped the Prudhoe Bay structure. On reading this years later, Harry Jamison was surprised because he said Humble had seen all of the Richfield's geophysics. In the real estate industry, and maybe the car market, this is called buyer's remorse.

This is just one of many ironic twists that occurred through the years. Humble is fortunate that they were not dismayed before they signed on, or they might have passed up their chance to participate in the Prudhoe discovery, yet to come.

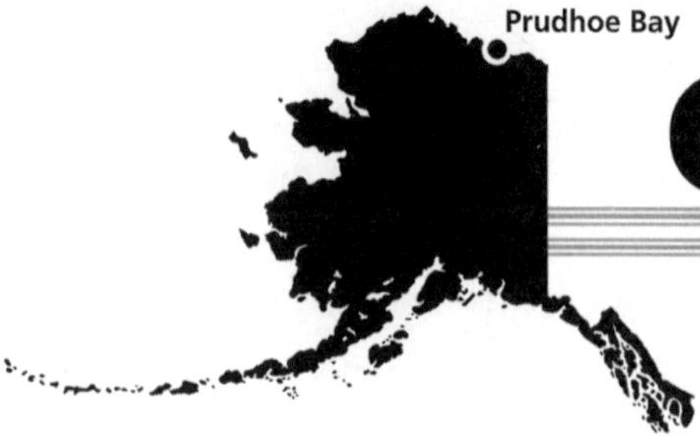

Chapter 7

The Atlantic Refining Company

In early 1961 the Atlantic Refining Company sent geologist Richard Crick and landsman Jerry Hill to Anchorage to learn as much as they could about the activities of the oil companies that were active in Alaska. They operated informally from their private residences. Their supervisors back at the Dallas, Texas headquarters decided, on the basis of what they reported, to open a formal exploration and land office in Anchorage.

One might have guessed that W. Dow Hamm, Vice President for Exploration and Production worldwide, would have had the authority to open a modest office of four people in Anchorage because he headed so large a division of the company. Not so; this humble beginning for Atlantic required the Operating Officer's Committee approval, which was the highest operating authority for the whole company. Needing this level of approval was an indication of the high degree of importance they placed on an immature and unfolding oil-producing province, recalls Julius Babisak, chief geologist.

> This meeting in Philadelphia [Atlantic Corporate headquarters] was to get the official sanction to place an office in Alaska, and be allowed to budget some real money. Dow Hamm, VP for Exploration and Production worldwide for

Atlantic, was with us. Our presentation was to Henderson Supplee, President and Chairman of the Board, Mr. Hamm, and the VPs of refining, marketing and finance, and all were somewhat cool to us Texans, I might add. [And Babisak might have added that those individuals, except for Mr. Hamm, were not at all familiar with the supply side of the business.]

Besides Louis Davis, VP of North American Exploration and Production, were Bill Moore, Manager of Exploration, Bill Campbell, Planning Manager, and me [Babisak]. Neither Louis nor Bill said much and Bill Campbell had the numbers they all liked to hear. [That is, the potential costs and rewards.] They really were given only as examples of what we might expect because there was no source of background experience to enable us to do otherwise.

I had worked up a presentation, which included what I had learned while in charge of the Army Special Mapping Unit in Northern Alaska during WWII. I also used USGS information obtained during the Naval Petroleum Reserve No. 4 exploration during the 1940s and early 1950s. The immense size of the North Slope was the point.

One of the few areas those Philadelphians had any respect for was Venezuela. After all, the ships from there brought oil right to their doorstep at nearby Fort Mifflin Refinery [on the Delaware River]. Cheap oil.

Using that theme, I had a pretty display of a large map of North America, on which I arced a rainbow across it from Venezuela to northern Alaska, with pots of black gold at each end, Yes, meaningless, but remember the people running out of heaven, to hell.

I also superimposed the state of California, spotting its oil fields on northern Alaska, suggesting how many fields there might be awaiting us. Bill Campbell and I agreed we did not know anything about a pipeline cost, but felt it shouldn't be a billion dollars, so I picked nine hundred million, and thinking of California and Venezuela, believed Alaska wells could be productive enough.

Supplee was intrigued. Hamm pushed us along, acting like we were telling almost as much as he knew. Can you see it? The rest were against it. What turned the day was when one of the locals, [refining and marketing members] complained that it would be very difficult to get the oil to Fort Mifflin. Supplee, in turn, was agitated and really put down Mr. Refining by ridiculing him for thinking that was the plan. In fact, at that time he turned to Hamm and announced that the project was on. No vote was taken. I was told that Supplee rarely had done that on an important issue.

It wasn't long before the nine hundred million was revised. I caught some flak for being presumptuous. That was true, but really I just wanted to go back to Alaska.

Julius Babisak mentioned Mr. Supplee's uncharacteristic decisiveness; it may not have hurt that as an ardent fly fisherman he made several trips to Alaska after Atlantic opened the office in Anchorage in 1962. The author unknowingly entered the picture in the fall of 1961, probably just after the Dallas people obtained the approval of the Operating Officer's Committee. During the fall budget conference, but at an associated social gathering in Calgary, Alberta where I was working then, Bill Moore, the North American exploration manager, put a hand on my shoulder and with a somewhat threatening tone said, "Boy, I'm going to send you to Alaska." That was about the same routine he had used a little over a year before when I was in Casper, Wyoming, and he threatened to send me to Canada. My reply was similar. "That's okay with me; when do I leave?" A college recruiter from another oil company suggested during an interview that was a good way to react. I was always able to do this sincerely because of the interesting places Atlantic sent me. The question about leaving went unanswered that evening, but it was a couple of months later, in early 1962 when the decision was made to open the Anchorage office, that I was sent with two others.

Atlantic opened the office in February in the Denali Theater building on Fourth Avenue in Anchorage. We were on the third floor. The long dimension of the building was oriented in a north–south direction and was essentially the width of one office wide with entry to the offices from a hall, which extended from the front to the rear of the building on the west side. My office was at the north end overlooking Knik Arm of Cook Inlet, with the port buildings and the Alaska Railroad yards in the foreground, and on clear days Mt. McKinley was visible over 100 miles away. Late in the day through the winter the peak became a pink shade, which was only ever captured in paintings by one artist, Sydney Lawrence. By today's standards the offices were rather spartan, but for then about as good as Anchorage had to offer and comfortable enough.

At the fall budget conference in 1962, our very first year there, we recommended a geophysical exploration program for the North Slope where Atlantic had already purchased 230,000 acres of oil and gas leases. Atlantic was thus an early entrant in the northern Alaska oil exploration.

By early 1964 our office staff had grown to ten or eleven, outgrowing the Denali building, and had moved into a new building a month before the Good Friday earthquake of 1964. The intensity of that earthquake, originally said by the experts to be 8.4 on the Richter scale, was later calculated to have been 9.1, which makes it the second strongest in the world. Since 1957, Alaska has had three earthquakes; the other two were the third and sixth most powerful in the world.

How fortunate we were to have moved out of that building. It was one of the most photographed buildings that made the rounds of national news, with its marquee at street level. It had not fallen off; the building had sunk that much because it was on what became known as the Fourth Avenue fault. Before the earthquake there was a crack almost an inch wide in the floor, on the third floor level of the stairwell, which was in the middle of the building. The building broke in half down the middle, and since then there has been little doubt the crack is where the building must have broken, or certainly very close to it. The crack almost certainly had to have been a manifestation of some prior earth movement of sufficient magnitude to have caused a break that large. Except for sinking and breaking in half, the building was strongly constructed of concrete and reinforced, so the two halves remained intact and were not devastated. Had we been working, the potential consequences are sobering to contemplate. Not only had the Atlantic office been moved, but also Good Friday was a company holiday.

Our friendly competitor, Mobil Oil, also had offices on the second floor of the Denali Building, but they did not have a holiday. Their office was small with only Bill Cook, the landman, and his secretary, Peggy, as the staff, and they were still working in the late afternoon when the quake struck. The quake lasted more than four minutes and thirty seconds, which Cook and Peggy rode out in a vault part of their office suite as the building sunk and disintegrated around them — the noise was horrendous, the shaking incessant. Bill and his secretary made it through without injury, and the seriousness of their situation was soon forgotten. Bill suffered much good-natured kidding about spending the time in the vault with his secretary, but they were terrified all the while.

Atlantic had relocated to the second floor of a new office building above a branch of the Matanuska Bank on the southwest corner of C Street and 8th Avenue. As noted, none of the Atlantic group was working because of the Good Friday holiday. After the quake, the office was a shambles — windows broken, curtains flapping in the breeze, file cabinets on their sides with many of the contents strewn across the floor.

Of course, there was no telephone communication to the outside. Two days later, Sunday, no one knew when the phone could be restored, so we dispatched Dick Crick, a bachelor, to Seattle to call all of our families and our Dallas supervisors and tell all that everyone was safe and unharmed.

Julius Babisak in Dallas remembered years later, "It was very rewarding to be the messenger of good news, especially to Gordon Davis' [Atlantic's Anchorage landman] parents in Dime Box, Texas."

None of the Atlantic employees were injured or their homes significantly damaged. The new office was devastated, and had we been working no doubt there would have been bumps, scrapes and cuts, but in all likelihood nothing serious.

Milton Norton and his family were not so lucky. This was before the Atlantic and Richfield merger to make ARCO, so Milt was a Richfield employee. The Nortons had been in Alaska since 1962 when Milt was sent to the Kenai Peninsula, a hundred or so miles south of Anchorage, to report on the geology of several exploratory wells Richfield was drilling. Glenda Norton relates some of her experiences there.

> Milt...invited me to go for the summer. I had visions of washing in the river for Yvonne, age almost 2, and Mike 5. I told him I would stay for a week. The geologist trailer was a new mobile home with two bedrooms, hot and cold running water, and a gas cooking stove — very nice. It was on the Anchor River at the town of Anchor Point. I stayed for the summer. The local Laundromat was one coin-operated washer in the "basement" (more a dirt cave) under the motel. The washer and two wooden boxes were the only things there. Bare light bulb and washer plugged into the outlet on a 2x4 on the

ceiling. Yvonne would scream every time we started down the stairs. Each time it started to spin, I put her down, put my foot in the coin return spot at the bottom of the machine and laid over top of the machine so it didn't spin itself out and unplug. The dryer was a bag of clothespins for the clothesline outside. Beautiful view of the river from the clothesline.

Later, in 1964, the Nortons lived in the Turnagain-by-the-Sea subdivision, which almost became Turnagain *in* the sea, during the earthquake. Before that, Turnagain by the Sea had a reputation as the most upscale housing area in Anchorage with some notable residents such as Lowell Thomas Jr., son of the longtime radio newsman. This residential enclave had the greatest destruction and loss of life during the earthquake.

The Turnagain homes were on a low bluff overlooking Turnagain Arm of the Cook Inlet, and it suffered a series of horizontal landslides that slithered to the northwest into the Turnagain Arm. These slides took the shape of numerous stair steps from the more stable, but badly cracked area, farthest back from Turnagain Arm down into the waters of the Arm. Many of the homes in these "stair steps," were in a wide range of destruction and none of them was salvageable. Here is Glenda's account of the events.

On Friday March 27, 1964, I had gone to the grocery store after my dentist appointment. When Milt got home from the office about 5:10 I had the grocery bags on the table. Had put Easter eggs on to boil. Mike was in the back yard feeding his sled dog, Maggie.

We were not alarmed when we felt the first tremor at 5:36. We had grown up with earthquakes in California, and had had a couple of minor ones since we moved to Anchorage two years ago. When the tremor didn't stop after a few seconds, I stepped over and turned off the burner under the eggs. As the shaking intensified, Milt picked up 3½-year-old Yvonne and we opened the back door and stood in the doorway. Standard quake instructions, stand in a doorway, go into a bathroom, or crawl under a desk. From that doorway we could look down into the backyard and watch 6-year-old Mike. By then he had fallen to his knees and was trying to get up. The power lines over his head were pulling and stretching as the house pitched and rocked. We were holding on to the doorjamb to remain standing.

The cans, bottles and jars among the groceries were crashing to the floor. Other things were hitting the floor, stereo equipment, books, fish bowl, etc. The Fostoria glassware coming out of the hutch was shattering around our feet. I remember thinking, "I won't have to worry about those next time we move." Our focus was still those electrical wires dancing over Mike's head.

I often hear people say the sound they remember after an earthquake is the sound of a freight train coming. Do you know the sound a nail makes when you pull it out of a board with a claw hammer? That is the sound I remember. Thousands of nails screeching in and out as the house shifted back and forth, back and forth, never ending, at least for the four and a half minute duration.

There was 12 to 18 inches of snow on the ground. As we held on to the doorjamb and looked down into the back yard, we saw a crack about three inches wide. It came from under the back fence, slithered across the yard, like a black snake, right past Mike, and hit the house. It felt like a truck had hit the house. It broke the cement block wall and twisted and broke the concrete foundation.

As the violent pitching of the house slowed, Mike was rolling around in the snow and couldn't get up. We held on to the walls and made our way to the hall closet to get our coats. Made our way down the back stairs as the house was still moving, and got the children in the car. The emergency whine was on all the radio stations before they announced what was already abundantly obvious to us. A neighbor came running out of her house with three of her children, after she saw a two-story house behind her house disappear. The Williamson's house, another two doors from hers slid over the cliff in slow motion. As we drove away, the street was still undulating in waves about three feet high and all the trees were pitching and waving.

Milt had told us to meet later at our friends' (the Murrays) house as he and neighbor Preston Locher were off to look for people who needed help. The slide area was about two miles long, but the housing area affected was two blocks wide and as much as six blocks inland. It was fortunate more of the slide area had not been developed. There were 75 houses in Turnagain destroyed. Many people were not home, and those who were rode out the landslide. There were only four fatalities, and those were four blocks west of us.

Preston and Milt climbed up and down the 40- to 50-foot crevasses of dirt in the slide roped together. Milt crawled along inside one neighbor's house where the ceiling was only four feet above the floor when a strong aftershock of rocking gave him a strong wave of vertigo, so he followed the rope back to the outside. After that they called out for people and knocked on walls. They came to a house with the carport completely buried with just the grill of a car visible. The house was in three or four pieces and when they called and knocked on one, a window opened. It was a second floor bathroom window, now

> 5 feet off the ground and facing the sky. A young woman poked her head out. She and her children were upstairs when the house started shaking. As the shaking grew more intense, she had gotten the three small children into the bathtub, the only place one could hold on to three small children while the house was breaking apart. Milt and Preston got them out and helped them up over the fault blocks to a safe place. After they had searched the last house it was dark. Milt says all he remembers aside from the exhausting climb up and down each fault block, was the eerie silence, then with the sound of shifting soil, the creaking and crumbling of broken houses with each aftershock.
>
> At Murrays' house that night, we had a fire in the fireplace for heat. It was not too effective because of the three and four-inch cracks on each side. At 5 the next morning a big aftershock took down a couple more houses on the edge of the slide area.

The next morning their drinks were made from melted snow because the water system was ruined. They drove back home and discovered a four-inch crack along the foundation of the house, so the Nortons quickly got as many essentials as possible in a few minutes. Subsequently, the Nortons' and their neighbors' homes were condemned, so with the help of fellow Richfield employees they moved the Norton household goods into storage.

On the day of the earthquake I had flown to the Kenai, about fifty or sixty miles south of Anchorage, with our oldest son to look for a cabin that we might move to a recreational lot we had overlooking the Cook Inlet. While waiting for the return flight I had stopped at a drug store for something, and as I backed out of the parking space there was a Jeep parked on my left with a raised snow plow. I noticed the snowplow was moving back and forth, but I had not felt anything, so immediately wondered if I had bumped into the Jeep and made the snow plow wiggle. In another split second I realized I was not close to the Jeep, and then we began to feel the motion. There were many ten- to twenty-foot-tall spruce trees close by whipping back and forth rhythmically, and the shelves in the drug store were unloading with the same rhythm. When the quake seemed like it would go forever, I decided to time it. I recorded over four minutes and the official length was four and a half minutes. Kenai was a small town with not many buildings, so we did not see or sense much devastation, but we soon discovered how serious it was when we went to the airport to return to Anchorage. We were told there would be no flights because the entire Anchorage infrastructure was out of commission.

We checked into a motel and the owner happened to be an amateur radio operator. The reports coming over the amateur bands were of the incomprehensible devastation. I never went to bed because it was the most worrisome night I ever spent in my life. I knew our other four children and two foster kids would be watching the *Mickey Mouse Club, Lassie* and their other favorite late afternoon TV shows. We had the TV room in the basement, which had two large glue lam beams supported by a post near the center of the room. The room was arranged in a way that some of the children would be under those beams. All night I could not dispel the vision of those huge beams falling on the children.

The next day at the airport no one had any idea about when scheduled flights would resume. Another man from Anchorage (a stranger) and I wanted so desperately to get to our families we shared the cost of a small plane charter flight back to Anchorage. The sky, with ragged gray relatively low overcast, didn't look inviting. Over Turnagain Arm the ceiling decreased until it met the water on the north side of the Arm, and we could not see the north shore. The pilot didn't want to fly over Turnagain Arm because there was not sufficient ceiling to gain the altitude to guide to land in the event of an engine failure. The clouds and fog made both ceiling and visibility too close to zero. He was reluctant, but said he would go if we insisted. We were desperate to get to our families, so he relented. He never should have accommodated us by breaking a supposedly inviolate safety rule. It was white-knuckle time over Turnagain Arm until we finally sighted the Anchorage coastline. Landfall should have been in the area of the Anchorage International Airport directly across the Arm from where we left the Kenai Peninsula, but we had drifted far off course, five or six miles inland and not far from the Chugach Mountains. After getting reoriented over land, we flew directly over our house, and there was no obvious damage. Several years later I saw that pilot, and told him who I was. His first words were, “Boy, I’ll never do that again!”

We hitchhiked home and found the only damage was a large Japanese glass fishing float that fell into a pot of stew my wife was making for dinner. Everything else, and all the family, was safe. The rest of the family were as relieved to see us as we were to find them safe. They had no clue about our status or whereabouts; we at least had known that they were home, regardless of my overworked imagination about the dangerous possible consequences of the quake. Marv Mangus had come by to check my family and advised my wife not to build a fire in the fireplace because of possible gas leaks, so everybody slept with their clothes on in sleeping bags in the livingroom, in the event of an immediate evacuation.

The Turnagain area slid out into the Knik Arm because the land there was mostly fine silt and sand, and it was underlain by a zone that had a high water content. The shaking of the earthquake caused this very wet zone to become jelly-like. The bluff along Knik Arm had nothing to hold it back, so it slumped and slid to a lower position. This created a domino effect that caused the whole Turnagain area to continue to slide into Knik Arm until all the land had reached a point where it finally restabilized itself. The reason there was little or no structural damage in the area in which we lived was because our subdivision had been built in an area of an extensive gravel bar.

All the Atlantic people attempted to go back to work the

following Monday. In retrospect, it was a futile effort. We should have stayed home because after picking up the file cabinets and replacing the contents there was nothing to do. The windows needed repair, as did all the windows in town, and the telephones didn't work; there was no heat and the plumbing was uncertain. Worse, everyone was in a confused state of mind, and there was the constant concern about potential aftershocks. It was a very unusual feeling, and something none of us had ever experienced. I had the constant feeling I should be doing "something," but had no idea what that something might be. What was it, this feeling? Was it an emotional letdown? I have wondered if a couple of weeks of complete relaxation back home, or at a resort might not have been good for everyone. There were no answers then, or now. It was several weeks before we were back to a normal routine.

Our kids played "earthquake" for weeks afterwards. As they would be playing someone might shout "earthquake!" whereupon they would all huddle together in some seemingly secure place like a corner or under a table. Shortly afterward they would resume what they had been doing, as though nothing had happened.

Finally, things at work got back to normal as windows were replaced and the heat was turned back on. The telephones worked, and the toilets flushed. We were back in the oil-finding mode.

During this era, some other oil companies had been more quick to move into Cook Inlet after the Swanson River discovery than Atlantic Refining had been. They had made significant discoveries in Cook Inlet on Middle Ground Shoal. This stimulated much geophysical work in the Inlet, and Atlantic was an active participant in Cook Inlet exploration and also on the Arctic Coastal Plain. Atlantic had prospects and leases in Cook Inlet before the merger with Richfield. These were drilled by ARCO, following the merger at Trading Bay; following that, ARCO drilled the *MacArthur River State* No.1 as a wildcat. Both produced oil from the Hemlock Sandstone the same producing zone at Swanson River. As it turned out, the MacArthur

The Atlantic Refining Company's first geological party field camp located just east of the Canning River.

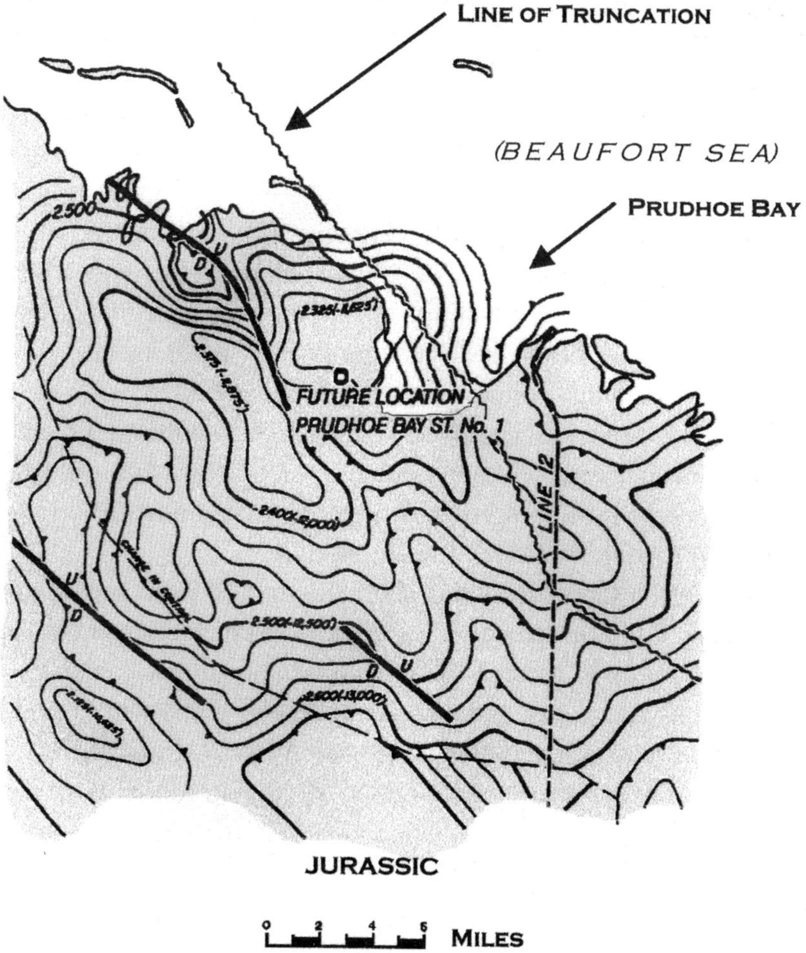

Figure 11: Atlantic Refining's seismic map for Sale #14, 1965. Atlantic's first map of Prudhoe by L.G. Pipes Contour Interval 125 feet. Used by Atlantic for 1965 state sale. Courtesy of Atlantic Richfield Company. After Specht, R.N. et al. Geophysics, *Vol. 51, No. 5 (May 1986) SEG (Copyright) 1986 reprinted by permission of the SEG whose permission is required for further use.*

River well was an extension of the MacArthur River field previously discovered by Union Oil. This field is the largest in Cook Inlet and has produced more than 600 million barrels of oil — far smaller than the future Prudhoe Bay, but still a national-class field.

The McArthur River discovery was overshadowed by the excitement over the Prudhoe Bay discovery, which became the only talk of the town, state and nation. MacArthur River oil field is historic in its own right because it had been probably more than fifteen years since any very large oil field had been discovered in the United States or Canada. That would be since the very large reef oil fields in West Texas and Alberta were found. It could be described as a giant discovery as compared to a 100-million barrel major field. In spite of its large size it was the last significant field discovered in the Cook Inlet. This happens frequently in oil exploration, the largest being found late in the exploration sequence. History suggests it was the largest single accumulation in which Atlantic Refining had ever significantly participated.

In 1963 Atlantic Refining began to supplement the North Slope geologic data from USGS geology dating back to NPR4 days, when Marvin Mangus returned to his former USGS haunts and was party chief. The first field camp was just east of what is now the western edge of the Arctic National Wildlife Range.

In 1964 Atlantic Refining shot the first seismic survey line across the Prudhoe structure. This was a reconnaissance line, so Lonnie Brantley, district geophysicist, and Jack Carlisle, a geophysical interpreter, were bowled off their feet when they saw the data because of the potential it showed for the area. They made the first Atlantic map interpretation of the Prudhoe Bay area. Concurrently, Larry Pipes on the Dallas staff was also making an interpretation (Figure 11). Pan American Petroleum and Sun Oil were partners for some of the geophysical work. Those companies lost interest and did not participate in bidding for leases at the Sale 14 by the Bureau of Land Management for Oil and Gas leases in

July 14, 1965. They are among the companies that missed out on purchasing the leases, which would ultimately produce oil.

Not all geologists in Anchorage at the time had great enthusiasm for the Arctic Coastal Plain as a budding oil province. One day at lunch a contemporary geologist with Union derided us in Atlantic Refining for exploring on the Arctic Slope. He said with great sarcasm, "Why you would have to find 500 or 600 million barrels of oil" [to pay for the development]. The implication was that it would be impossible to find a field that large. How wrong he was about that part, because even his estimate would not have been enough had that been all ARCO found. On the other hand, statistics were on his side. It was very unlikely we would find something large enough to be commercial, based on what was known and given the industry history and experience at that time.

Harry Jamison with Richfield heard essentially the same dismissive comments on a flight to Seattle from Anchorage. His seat mate, a senior member of Union Oil management, castigated him for being in a region that needed "a billion-barrel oil field" to be commercial. At that stage of the game, perhaps Union people were justifying in their own minds why they were not participating in the North Slope (Arctic Coast) game. Again, it was interesting that even the Union official's billion-barrel fields would not have been enough to justify development. The comments of these two Union oil employees are indicative of the scale of size most all oil people were thinking. No one that I know envisioned what was eventually found, so in that regard the Union folk were not alone.

Like all of the other companies active on the Arctic Slope, Atlantic Refining had essentially the same geophysical structure map as — and contemporaneous with — the rest of the companies active in the area. Specht, chief geophysicist, and Selman, district geophysicist, along with their associates Brown and Carlisle, have documented the geophysical efforts of the participants who later became part of ARCO in *Geophysical Case History, Prudhoe Bay Field Geophysics,* 1984.

The Atlantic Refining Company and Richfield Oil Company first talked about the potential of a joint operation in the fall of 1961. Julius Babisak, then Atlantic chief geologist, recalled the discussions.

> Four of us from Dallas went to Los Angeles to meet the Richfield people on a Friday. This suited me just fine. I stayed over because, on Saturday, Ohio State played Southern Cal in the Coliseum. We exploration people got along very well. We were anxious to see a joint agreement made that would lead to a partnership. The business people's discussion was OK, but they were forever fearful they might give something away. I reported that from an exploration standpoint I was ready to go. Our business representative, having been assured as to the exploration viability, proceeded to outline our position. He insisted that Atlantic would be the operator. This meant we would do the work and bill them, and keep them informed. This is done routinely, if the other party sees a good thing happening, or if they know and trust you.
>
> We had not really dealt with each other before, and they were offended. After a short recess they made the same statement as our man did, as their position. Our man says, "Meeting over, goodbye." [An individual such as this was once described as being able to strut while sitting down.]
>
> Atlantic Refining and Richfield could have been partners early on, without BP or Humble [now Exxon Mobil] in the picture. Just think, we could have each had half. And we would each be a big company, one in the west, the other in the east.

Pride in a company's operating ability was always a sticky subject when discussing joint operating agreements. The executives in each company always believed their respective company could operate more efficiently than the potential joint partner. Sometimes it was true, other times it was not. It was not unusual for this to be an issue on which neither company would give in. That was the situation between Richfield and Atlantic in the early joint operating area discussion. The Atlantic person in this situation was presumptuous because Richfield had a record of success and more than four years of operating experience in Alaska. Atlantic at the time had nothing in Alaska. The Richfield people no doubt read this as arrogance, bringing an end to the discussion.

Later the Atlantic Refining Company and Richfield Oil Corporation agreed in principal to merge on September 16, 1965. The merger detail discussions culminated in Richfield merging into the Atlantic Refining Company on January 3, 1966. A month later, on February 2, 1966, R.O. Anderson, chairman of the board of Atlantic, and E.M. "Mo" Benson, an executive with the former Richfield Corp., traveled to the North Slope together. Mo related the following.

> When Bob and I arrived at Fairbanks that day, the Richfield employees had gotten a freshly painted, green and white Atlantic Refining Company sign to hang on the office wall, covering up the Richfield name. When Anderson and I returned to Fairbanks [following a press conference], Bob said, "Mo we have to change the name of the company." I asked, "Why?" He said, "We have to get the Richfield name included, or every time I come to Alaska I will have to explain what the Chairman of the Board of Atlantic Refining Company is doing in Richfield's area." At the shareholders meeting on May 3, 1966 the name adopted was the one we use today, Atlantic Richfield Company, or ARCO.

There was not a great amount of overlap where the two companies operated. Atlantic was an east-coast company with all its marketing there and with production in the mid-continent. Richfield was a west-coast company with marketing and production mostly in California. There was significant overlap in Alaska, with each company having a full staff, which were successfully combined with little disruption of personnel. At the time of the merger the Atlantic Anchorage staff had grown considerably, with the result that ARCO's amalgamated workforce was approximately double that of the two former companies. Little did they know how many more people the discovery at Prudhoe Bay would add.

Oil leases are what give an oil company the legal authority to drill for oil, and the newly formed ARCO had to consolidate its leases and prepare for the future. Among the leases were some on the sleeping super giant at Prudhoe Bay brought to the merger mainly by Richfield and Humble; therefore, when "ARCO" is used from now on, readers should think "ARCO-Humble" (now Exxon Mobil). The following background on Humble will show something of that company's agonizing Alaska history and how, through the persistence of one man, they got a piece of the Prudhoe Bay pie.

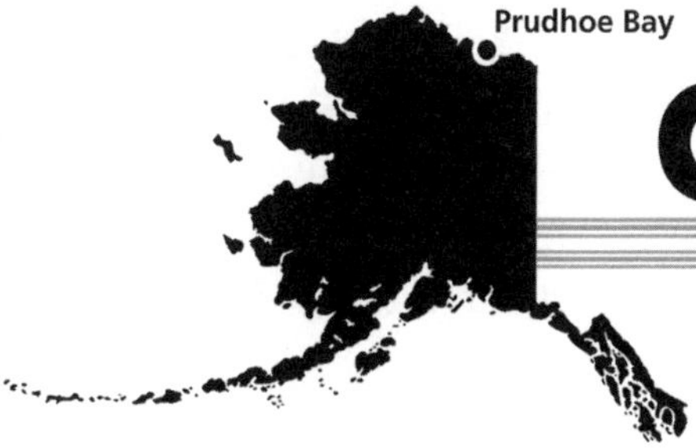

Chapter 8

Humble Oil Negotiates a Way In

In March 1959, Humble Oil and Refining Company (now Exxon Mobil) completed what was the most expensive dry hole it had ever drilled. This duster was located at the *Bear Creek Unit* No.1 on the Aleutian Peninsula. Shell Oil was Humble's partner on this project. Humble's share of the cost was $7 million (equivalent to $20.7 million in 2006). Costs have inflated so dramatically that $7 million doesn't sound like much now, but it was big money then, even for a company like Humble.

As an exploration and production company, Humble usually had a former exploration person (a geologist) as company president. At that time it was Morgan Davis. In spite of his geological background, the Bear Creek well soured him so much on Alaska that he closed the Anchorage office. This was the same company in which Wallace Pratt was the first chief geologist and industry philosopher, who warned that the geologist could fall into a trap. "His knowledge makes him over-conservative, or obscures for him the fact that much remains always unknown to him. If his knowledge blinds him to the unknown, his discoveries will be fewer."

J.R. Jackson was handling Humble's exploration affairs in Los Angeles. Fortunately for Humble, he was a believer in the future potential of Alaska. In 1960, a year after the Bear Creek disappointment, Jackson convinced John Loftis, general manager of Humble's Minerals Department, that Humble should at least do some geological evaluation of Alaska. Dean Morgridge, geologist, and Robert J. Walker were sent to Anchorage to do that and to evaluate oil industry activity.

Merrill W. Haas, vice president of exploration, supported J.R. Jackson in his enthusiasm for Alaska exploration. Morgridge and Walker thought they could assemble sufficient information to make it possible for Humble to be competitive. In May of 1963 Dean Morgridge made a detailed analysis of the prospects and recommended staffing, and a budget of just over $1.3 million for 1964 and 1965. The Humble parent company, Standard Oil of New Jersey, was not sufficiently impressed and refused to allocate sufficient funds for more exploration. This very well could have been from residual disappointment over the expensive Bear Creek dry hole.

Frequently, when a company is shut out of an area or wants to increase its representation, it will seek entry via a partnership with another company that has a lease position and an exploration background. One might wonder why a company with information, and presumably in a better competitive position, would agree to take on a partner. This may occur for any number of reasons, but there are two common motives that lead a company to accept partners. One is a lack of funds, and the other is a decrease in enthusiasm for a given area. Frequently, both reasons apply.

Humble's involvement should be credited to J.R. Jackson and his staff. Harry Jamison, then of Richfield, relates that Mason Hill, Richfield's chief geologist, contacted Jackson about a potential partnership. This was music to J.R. Jackson's ears because he was looking for an entry to North Slope exploration. This was a different idea than the one for which the Standard Oil parent company had refused funding. Harry then made a presentation to

J.R. Jackson, Dean Morgridge and Ken Fuller, a Humble Oil geophysicist. Richfield offered to let Humble in on everything they had done up to that point for $3 million, after which Richfield and Humble would share everything on a fifty-fifty basis. That amount of money for work already done in an area this remote was a substantial windfall for a company of Richfield's size. Likewise, having someone to share potential future expensive field exploration expenses also had great appeal to Richfield.

Mr. Jackson was more successful than he had been before by convincing his management that Humble should say yes. He negotiated the terms with Frank E. McPhillips, a land manager for Richfield, and a letter of intent was signed July 22, 1964. It covered twenty-five million acres, which is all of the Arctic Slope north of the Brooks Range to the Arctic Ocean and from the northwest coast of Alaska to Canada.

When this was presented to the Humble executive committee for approval they made the observation that from one to one-and-a-half billion barrels would be needed for a discovery to be successful. Humble believed it was paying a hard penalty for its entry. Buyer's remorse again. Subsequent seismic work by the partnership covered the Prudhoe Bay structure, and Charles Boles, a Humble seismic interpreter, made the first detailed contour map of the Prudhoe Bay structure. By making this move, Humble Oil was back in the mode that would have pleased its patriarch, Wallace Pratt, "so long as a single oil-finder remains with a mental vision of a new oil field…oil fields may continue to be discovered."

There are frequently "what ifs" in some ventures. In preparing for the July 14, 1965 lease sale at which the all-important tracts on the Prudhoe Bay structure were bought, Humble wanted to allocate another million dollars to be matched by Richfield. Richfield just could not do it. Had they been able to, the Humble–Richfield joint venture would have owned the whole structure.

The leases they would have bought were farther down structure and contained more of the oil column than the top leases the Richfield–Humble group actually bought. Classically, an oil field might have gas at the top or the highest elevation position. Below the gas might be oil, and below the oil, water. Prudhoe Bay was such a field, but before the lease sale no one knew, so, as was customary then, they placed the most value on bids for leases at the top of the structure. So for that two million they didn't spend, they missed getting perhaps as much as seven or eight billion barrels, maybe more.

After the discovery it was obvious that Humble had not paid the hard penalty their people had fretted about. It is anyone's guess what might have happened had Mr. Jackson succeeded in getting the budget for Humble to explore on its own. They almost certainly would have done the seismic work and without doubt they would have seen the structure. All the other companies did. Who knows what might have happened beyond that? Sadly, Mr. Jackson departed this life before he could see the credit I give him for Humble's success. There are no doubt others who think they have a share in this credit. They very well could be right, and they will know this in their hearts.

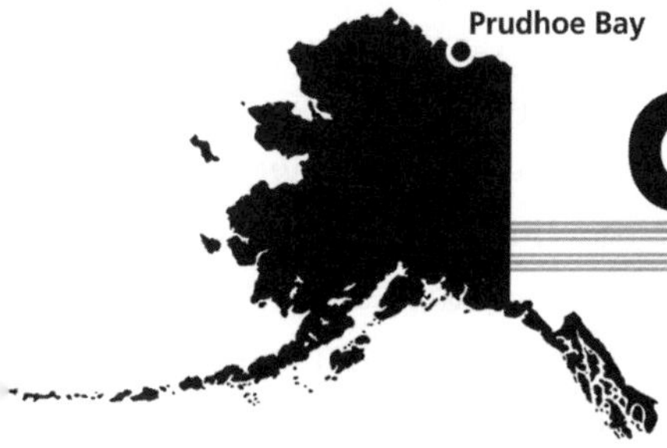

Chapter 9

Sinclair: Senior Management Blinked

There was another company with an Alaska staff that was also enthusiastic about the Arctic Alaska oil potential, but their top management blinked.

How many times have you heard how good it is to get into something early that turns out to be a bonanza? I think the current vernacular for that is "being ahead of the curve." Loren Ware, the domestic exploration manager for Sinclair, wheedled enough money out of the board of directors to put the first oil industry geological field party to work on the North Slope in 1958. Often the top people in corporate offices have a hard time relating to frontier oil exploration because it is, more often than not, fraught with disappointment.

During that first summer British Petroleum (BP) sought out the Sinclair people and asked for help getting started in Alaska. This is interesting from two angles. The first is that one of the largest and most successful international oil companies approached a medium-size U.S. company for advice. That was a feather in the hats of the Sinclair people in Alaska. The second angle was that these Sinclair geologists, who were held in high regard by a large competitor, were the ones who had to grovel to their management for funds.

Why, one may ask, would anyone want to help a competitor? This was common and a professional courtesy among oil industry exploration people. There always had been a fairly open exchange of general background information between companies early in the explorations cycle. At that stage usually what is known is already in the public information domain and thus is not proprietary. As companies build on this information base, they begin to acquire information that is in their best interest to keep to themselves. Then, help is usually on a *quid pro quo* basis because information is believed to be so valuable it is under strict security.

On occasion this early sharing of information leads to a bond of trust and mutual admiration between the helping company and the company receiving the help. This can easily flow into a formal joint exploration agreement. Sinclair and BP jointly became a major participant in North Slope exploration in 1959.

Loren Ware had obtained enough of a budget to begin acquiring lease blocks during 1959. Beyond that he had no budget for

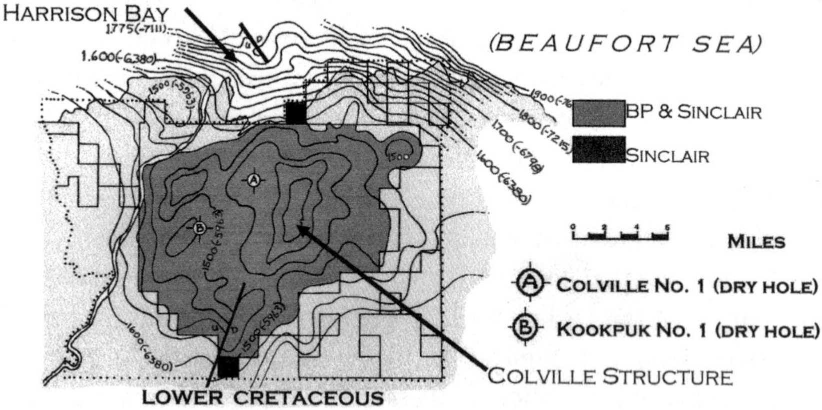

Figure 12: The BP and Sinclair follies. This is a very large geologic anticline, The Colville Structure, about 20 x 25 miles per side, or 500 governmental sections of 640 acres each for a total of 320,000 acres. The shape is almost a classical symmetrical dome. This structure was offered for competitive lease sale by the state of Alaska, before the Prudhoe Bay leases were offered. Because of its very large size and the fact that leases were offered, Loren Ware with Sinclair decided to spend their money in the shaded area within the structure.

After Specht, R.N. et al. Geophysics, *Vol. 51, No. 5 (May 1986) SEG (Copyright) 1986 reprinted by permission of the SEG whose permission is required for further use.*

drilling, as Sinclair's top management was not interested in Alaska. Wallace Pratt, the famous oil-finder discussed previously, had this to say relative to this situation: "to refrain from venturing because the evidence is not wholly favorable, is often near tragedy for the geologist." It was tragic for Sinclair.

The field parties in 1958 and the assembling of leases in 1959 made Sinclair the earliest industry player on the Arctic Slope of Alaska. In 1961 Sinclair joined Colorado Oil and Gas to drill a well on the Gubik surface anticline. It was completed as a small gas well. By the end of 1964 Sinclair had drilled two other dry holes on the Arctic Slope.

At the Anchorage level, Arthur L. Bowsher had stimulated Loren Ware's interest in northern Alaska. Bowsher, who had worked on NPR4 for the USGS and was now with Sinclair, reported that Sinclair and BP sent the first industry seismic crew to the area in 1962. Sinclair, and later BP, mapped the giant Colville Structure (Figure 12) and the Prudhoe Bay structure, as did all the other companies operating in the area. Figure 12 is G.L. Scott's map of the Colville structure. This structure is about 20 x 25 miles per side, or 500 governmental sections of 640 acres each, and total of 320,000 acres. This is the area favored by Sinclair's Manager for Alaska, Loren Ware, and where he chose to spend Sinclair's money. One of their top geophysicists, Paul Lyon, did the first interpretation of structure of Colville and Prudhoe. The Colville Structure was a classical-appearing potential oil trap, and that is where Sinclair and BP decided to cast their lot.

Mr. Bowsher opined that Loren Ware made the decision to go for the Colville leases in a major way because that was a fact situation, i.e., a known structure, and had the potential to get leases. They also knew about the Prudhoe Bay structure, but at that time they didn't know if or when the leases might be offered. This was straightforward and logical thinking. Prudhoe was large, but with ambiguity about part of the structure, and no time certain for leasing. As has occurred repeatedly in oil exploration, a prospect with less obvious merit — in this situation Prudhoe — ultimately became the big enchilada, the elephant, the super giant.

Sinclair and BP drilled the *Colville* No.1 and it was dry. A year later Union Oil drilled the *Kookpuk* No.1 on another part of the huge Colville Structure. It, too, was a dry hole.

When the Prudhoe Bay leases were finally offered for sale, Mr. Ware could not convince the corporate officers to provide bid money to participate, and shortly thereafter he was transferred to Houston as vice president of their Gulf Coast District. In the *Geophysical Case History, Prudhoe Bay Field,* Specht, et al. put it this way, "Sinclair elected to compete only on three tracts [on the Prudhoe Structure], due to lack of success in the Foothills play and their position on the larger Colville structure." Specht also pointed out another irony: "The final Sinclair map before State Sale No. 14 was drawn by G.L. Scott, and most nearly reflects the actual structure of all the present maps." Wallace Pratt, years before, had keyed on a dilemma. "What we learn, instead of illuminating what we have [or need] to learn, sometimes casts a mental shadow over it, rendering it less discernible and impelling us to ignore it."

Hearsay information had it that Sinclair missed a final opportunity. BP is said to have offered Sinclair the opportunity to participate in half of the future productive leases BP bought at Sale 14. Sinclair was said to have not even given BP a response.

Loren Ware had all the right instincts, was in on the ground floor, and even had the best seismic interpretation of the Prudhoe structure, but did not enjoy the fruits of the big play. It was left to others to justify his faith.

After the discovery at Prudhoe Bay, Sinclair drilled the *Ugnu* No.1, which was an oil discovery from the Kuparuk Sandstone. This same sandstone unit was also discovered to be productive in the Prudhoe Bay area and the two areas ultimately formed the Kuparuk field. Kuparuk has produced more than two billion barrels of oil. An irony was that Ugnu was on the Colville structure. A greater irony was the fact that little Sinclair helped the much larger international company British Petroleum get a start in Alaska, and ended up the dominant holder of the North Alaska oil production.

Walter Pratt's earlier comment about bad news obscuring the fact that there was much that remained unknown, was true in this case, and this was what infected Sinclair senior executive managers at the top. They should have supported Loren Ware, their man in Anchorage. Later still, Sinclair was bought by ARCO.

On to Prudhoe Bay…

Chapter 10

Leases for Sale

If an oil company does not own the leases, it doesn't matter that they have an excellent geological structure. Here is how this part of the picture unfolded. Opening the area east of NPR4 to oil and gas leasing was of vital importance to all concerned — the federal government, the state of Alaska and the oil industry.

In 1958 it was all federal land. That year the federal government had opened leases for simultaneous filing in the area east of NPR4 in Arctic Alaska (Figure 13). Simultaneous filing means that any company or individual may file an application for a lease within a designated area for a specified sixty-day period. In 1958 the tracts were four governmental sections of 640 acres each, or 2,560 acres. This was ideal for the "little guys" because anyone with the ten-dollar filing fee could participate. Those awarded tracts had to also pay the first annual rental of fifty cents an acre. Thus for $1,270 you could get a federal oil and gas lease — if you were the only applicant. The lease offerings were well publicized by the government, being first advertised in the U.S. Federal Register.

There were four simultaneous filings on the North Slope. It is not surprising with this low entry fee that everybody got into the act. One tract had 300 applicants. In this area anyone who was the lone filer was out on the proverbial limb because it was obvious no one else thought the tract was worth anything. The other side of that coin occurs when an individual's tract is beside the tract of an oil company. It has happened that more than one company was interested in an individual's tract. A lucky applicant was virtually guaranteed to get a windfall offer from the oil company. If that oil company would pay five dollars per acre it would yield an $11,500 profit. At the very first filing one individual working on behalf of BP had his name drawn for 55,000 acres. The size of that windfall was not divulged. At a later filing, Jack Roderick, author of an overall history of Alaska oil exploration, *Crude Dreams,* netted $50,000 on a $25,000 gamble. There was only one other applicant on one of his tracts, a Roman Catholic bishop from Fairbanks. He was not an oil company so Jack was sweating, but two companies eventually did take his leases off his hands (Figure 11).

In the case of tracts on which there were multiple applications, there was a lottery drawing to determine which applicants were the first, second and third filers for each of the tracts. Three names were drawn to facilitate the outcome in the event successive applicants did not qualify. This usually would produce a winner, but in the event it didn't, the process could go to another lottery drawing.

The federal land managers did not offer leases over the entire area between NPR4 on the west and the Arctic National Wildlife Range on the east. They omitted a wedge-shaped area on the northern coast between the federally leased lands just mentioned and the Arctic Ocean that contained over 1.6 million acres because it was State of Alaska acreage.

In 1960, Thomas R. Marshall Jr., a geologist, was named state land selection officer in the Department of Natural Resources. How wonderfully fortuitous this appointment was for the state of Alaska. He recommended that the state select the 1.6+ million acres mentioned above. His recommendation was frustrated by bureaucratic delay. The delay was understandable because it is easy to forget that at that stage of knowledge there was no evidence that oil would be found. The only positive thing was that many oil companies were doing seismograph exploration. Marshall continued pressing the recommendation until 1963 when the state formally selected all of the 1.6 million acres as part of the state's allotment under the Alaska Statehood Act. By this time Tom had been promoted to state petroleum geologist.

The state held the first state lease sale in 1964, and Union — and Sinclair with BP as a partner — bought the leases mentioned in the chapter on the Colville Structure. They paid about $32 per acre. At a state sale in July 1965, Richfield Oil Company, with partner Humble, and British Petroleum, bought the controlling leases on the Prudhoe Bay structure.

These sales were competitive sealed bid sales. They were, and

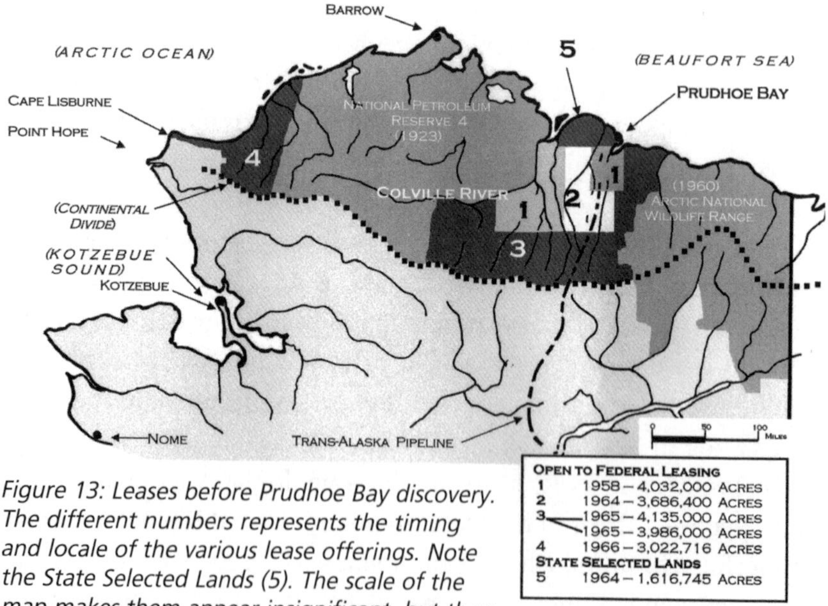

Figure 13: Leases before Prudhoe Bay discovery. The different numbers represents the timing and locale of the various lease offerings. Note the State Selected Lands (5). The scale of the map makes them appear insignificant, but they cover 1.6 million acres and all of the oil fields.

After Jamison, H.C. et al. Giant Oil and Gas Fields of the Decade 1968–1978. *AAPG (copyright) 1980 reprinted by permission of the AAPG whose permission is required for further use.*

still are, probably the highest stakes guessing games one would ever encounter. One large factor winnowed the field of participants, and that was the price each bid on a given tract. The tracts were the same size as in the simultaneous filing area — 2,560 aces. At that stage of exploration there were no parameters that would reveal a value of the tracts. The companies simply had to decide subjectively what price they were *willing* to bid on each tract, uncertain about the price each of the competitors was willing to pay. Lacking value-setting parameters, bidding became a matter of judgment, experience, guessing and "gut" feel. The questions permeating the discussions were: What will it take to get a given lease tract? What are we willing to pay for a given tract? How much total exposure can we afford; that is, if we bought every lease we bid on, what would we spend and can we afford it? What do we think we will actually spend? (Likely a lower number than what we can afford because historically a company will not win all the tracts.) How do you think competitors X, Y and Z are answering these same sorts of questions? What were the prices for similar acreage at other sales?

The exploration people participated in this exercise by rating the relative importance of each tract based on their perception of the producing potential. For the 1965 sale, Charles Selman, Richfield's district geophysicist, remembered that at the district level in Anchorage he, with Harry Jamison, thought a bid per acre about three times the $32 per acre paid on the Colville structure was about right, and that is what they recommended to their supervisors in Los Angeles. Bidding was mostly the province of land people because it was within their knowledge and experience. Top management also had to agree, because the dollar amounts were usually very large and had to fit in with the overall annual business plan.

In sealed bid sales there is the opportunity to just barely win a good tract or significantly outbid the next highest bidder. The latter is referred to as "leaving a lot of money on the table." In most sealed bid sales there are some companies that end up looking foolish, some looking wise, and still others appearing foolish on some bids and wise on others.

Apart from the usual guessing game all the companies had to play, Richfield had an additional unique internal problem through

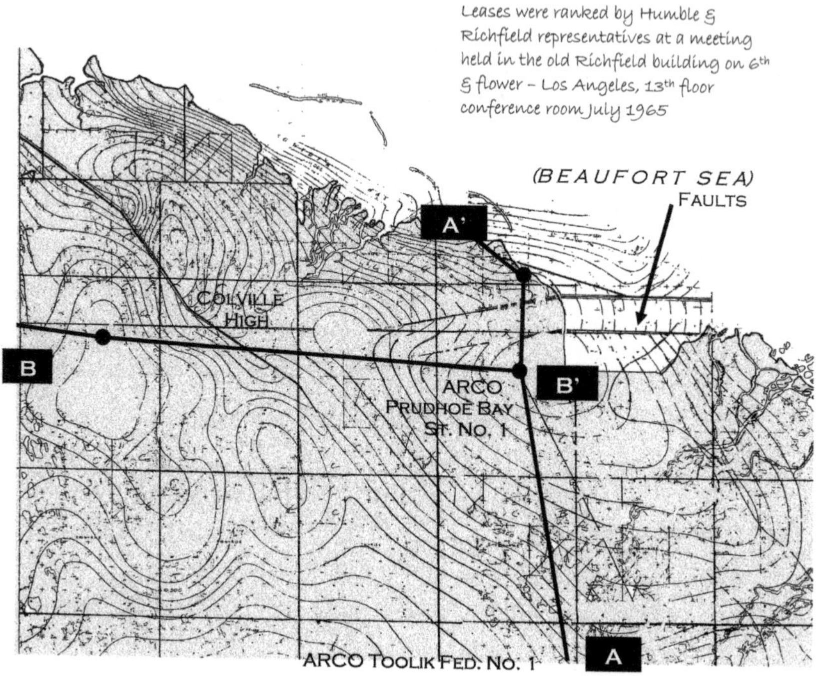

Figure 14: Richfield-Humble Seis Map. Contour Interval 100 feet. This is a copy of the map used by Richfield and Humble when deciding what to bid at Sale #14, July 14, 1965. The map closely represents the geological structure interpretation on which the Prudhoe Bay State *No.1 was drilled. Subsequent drilling changed only the details of this original structural interpretation from Reflection Geophysical techniques. The location of the lines of cross sections A–A' and B–B' (Figures 15 and 16) appear on this map and also on Figure 21. A couple of the faults and the truncation of rock units are labeled on the cross sections. A cross section is an idealized depiction of a vertical plane through the earth, then moving one side of the earth away so one is looking into the other exposed side of the plane. Maps and cross sections reduce large features in the earth to inches on paper and provide a somewhat exaggerated representation. The contour interval of Figure 14 is 100 feet. The contours are so close it might give one the idea of a steeply sloping rock surface. This slope, or dip as geologists refer to it, is 300 feet in 1.75 miles. This equates to a 3 percent slope. Most steep mountain highway grades are not more than 8 percent.*

much of its history. It began in 1937 when the chairman of the board and the president were convicted and sent to prison for stealing from the stockholders. Richfield went into receivership. Sinclair Oil already had an interest in Richfield by virtue of buying half of Richfield's predecessor, Rio Grand Oil Company. Cities Service also became part owner during this troubled period. Richfield's problem, therefore, was that its board of directors had members from Sinclair and Cities, who were also competitors. These competitors might be potentially troublesome when it came time to bid for the leases if, as "insiders," they knew what Richfield was going to do and used that knowledge against Richfield at a lease sale.

Rollin Eckis, Richfield's president and a geologist, always had great enthusiasm for the Arctic Coastal Plain. Now he also had seismic findings that revealed the Prudhoe Bay structure and its potential. Sharing this with the board of directors would be giving away the "family jewels." So, when Humble came knocking regarding a joint operation with Richfield, Eckis could share his enthusiasm with a partner rather than with competitors. Together, Richfield and Humble won big at the state lease sale when they bought the Prudhoe Bay leases. Eckis' memory of his visit to northern Alaska as a guest of the Navy during the NPR4 exploration never dimmed. "It was then and there that I developed the feeling that there was a tremendous potential for oil on the North Slope."

On Sale No. 14, July 14, 1965, Richfield and Humble were thought by some to be foolish (Figure 14). They bid \$93.78 per acre, nearly double the nearest competitor, British Petroleum, who bid \$47.60 per acre, and a little less than Selman and Harry Jamison recommended. The perception was that Richfield–Humble had left a lot of money on the table. Mobil was third highest, in the low double-digit dollars, and the Atlantic Refining Company a very distant fourth in the single-digit dollars.

When a company is out in front like this, it causes a certain amount of consternation, generating derision, envy, curiosity and questions about sanity. "How did they get to those numbers? They need their heads examined! How can they justify those prices for 'ram pasture'? What did they know that we didn't? They are going to have fun explaining that to their board of directors! I wish we had gotten a few of those tracts." I guess even some of the people in Richfield and Humble might have had a twinge of buyer's remorse about bidding over $45 per acre more than was necessary.

This sale provided much grist for the Anchorage conversation mill. But, oh how well it turned out for ARCO and Humble. It didn't become apparent for three years, but for the second time the rather small Richfield Oil Corporation had started what was to become a highly dramatic part of Alaska's history. There can be speculation about what might have happened had someone else had the winning bid, but the record is clear — Richfield did it. What happened was a result of Richfield and Humble looking at their data, considering the imponderables, and ultimately "sucking it up" and putting their money down. Their people were among the strongest links in the chain leading to the discovery.

British Petroleum looked like an also-ran, ending up with fringe leases. History would tell that Richfield bought what largely turned out to be the gas cap and British Petroleum ended up with the "oil ring." In a classical situation, when looking at a map view an uncomplicated anticlinal (i.e., not complicated by faulting) oil structure will frequently have gas at the top of the structure or the middle of the field. Below the gas there is oil and in map view that would be a band of oil (the "oil ring") in a map view, surrounding the gas. Below the oil is water, and the map view for that would be water surrounding the oil. ARCO–Humble ended up with lots of oil, but also a big share of gas for which a market has never materialized.

Tom Marshall deserves a monument for persisting in his

recommendation for the state to select the acreage where the Prudhoe Bay oil discovery was made. Tom is modest and would likely shrink from such recognition. The unbelievable dividends the state and individual Alaskans have enjoyed since the discovery have emanated from his appointment as the lands selection officer in 1960. How about a Thomas R. Marshall Chair of Geology at the University of Alaska Fairbanks?

Chapter 11

Oil Exploration: Spirit and Gumption

Oil industry geologists, geophysicists, or collectively, "explorers," are unique and operate with different thought processes than many other professions. They tend to function with a high rate of failure, because only a small percentage of exploration drilling projects ever find oil. Over a lifetime, an oil explorer would consider himself fortunate if a small percentage of projects uncovered oil or gas. Living with a low rate of success means swallowing some frustration. Yet the oil explorer must be buoyant, diligent and enterprising if he or she hopes to obtain management funding for drilling prospects.

Readers might logically wonder how an oil explorer can possibly maintain a sense of optimism and enthusiasm in the face of the low odds. The necessary ingredient is a sense of adventure. Lisa Rossbacher, a geologist who is also the president of Southern Polytechnic State University in Marietta, Georgia, sums it up:

> Geologists first ask "Why?" and then ask "What?" We think and speak in metaphors and analogies. We use our hands a lot when explaining ideas. We tend to look for the big picture and find the broader patterns in the detailed data. We're willing to take risks to try to put information together in new ways.

Adventure and excitement go hand in hand. They are ever present as one combines different facets of earth science to arrive at a compelling interpretation leading to the drilling of an exploratory well.

In the early days of the oil industry, explorers found satisfaction in finding and mapping huge anticlines on the earth's surface. (Anticlines are places where dynamic forces in the earth have caused flat lying rocks to be formed into an arch.) Anticlines were the primary targets for exploratory wildcat wells because many contained oil or gas. The earliest were nicknamed "sheepherder" anticlines because spotting them was so easy that even Wyoming sheepherders could see them from far off. From that point on, oil prospects became increasingly more elusive and now involve sophisticated seismograph field techniques followed by intricate computer manipulation and analysis.

Well-site geologists have two potential opportunities for adventure. One is watching the rock cuttings for changes in the formations and looking for the formation in which the oil is expected. The other, and more exciting, adventure is to see indications of oil in the rock samples. "Well sitting" is the jargon for a geologist "watching" a drilling well (i.e. describing the rock chips). I have fond memories of my well-sitting days in West Texas, where I experienced these two levels of adventure and excitement.

Sparks of optimism and fundamental geologic analysis fuel this overpowering desire to decipher the unknown. It is kindled by an unquenchable spirit and culminates in the drilling of an oil prospect. The following story paints the picture. A little boy was taken by his parents to his grandfather's farm. Later, the parents missed him, and they found him digging furiously in a huge pile of horse manure. When asked what he was doing he said, "With this much manure there surely must be a pony in there somewhere." The sense of resolve in this anecdote is reminiscent of the geologists in the going-to-hell story to look for oil, in the Richfield chapter. Some visionaries saw the North Slope of Alaska as such a place. ARCO District Manager H.C. Jamison described it this way:

> …being up there [in Alaska] in a remote location, all you know, most geologists, most exploration people, have that [adventuresome spirit] in their hearts someplace or they wouldn't be in the business in the first place. Certainly that was a very adventuresome experience over a number of years. I guess I'm...if not an adventurer physically, then I am mentally...there's the underlying technical drama...This occurs on a daily basis in looking for the unknown...most of our efforts are fraught with failure rather than success, but in all sincerity I love being in exploration.

Wallace Pratt, after many exploration successes, reflected on many of these same qualities. He characterized a good oil-finder in this manner: "The qualities which mark the individual oil-finder are faith, persistence, the venture spirit, and vision. If he is informed and trained in the art of oil-finding, so much the better." Notice that he puts the importance of character traits ahead of training. Definitely not blinded by narrow vision, in 1944 Pratt predicted the discovery of important oil deposits in the North Slope of Alaska. He was endowed with such an all-encompassing curiosity that he discussed the eventual use of solar energy long before its present promoters were even born.

The thirst for exploring the unknown motivated the Richfield Oil Company people to discover the Swanson River oil field on the Kenai Peninsula in 1957. It should be no surprise that the Richfield corporate culture led them to the North Slope of Alaska. Meanwhile, some other high-level managements thought, "They are going to hell," for exploring in so remote an area. Richfield sent its first geological field party to the Arctic Coastal Plain in the summer of 1959. By the early 1960s, nearly all of the companies with Anchorage offices (and some outside Alaska) had surface geologic parties working on the North Slope. With the advent of helicopters, the geologists' mobility was further enhanced, making it possible to refine the interpretation of the surface geology and bring back samples that could be evaluated for age, oil source potential and reservoir quality.

The Coastal Plain is devoid of rock outcrops, so these early geologists worked in the hills where there were rock outcrops. In light of the spotty distribution of outcrops, the geologists would piece the geology together bit by bit, and during subsequent field seasons attempt to add new information where there was none or where available data were confusing.

The cooperation of the companies mentioned before continued somewhat in this period, but mostly in operational logistics. The companies by now were beginning to accumulate data that their explorationists believed had competitive value, so informational cooperation fell by the way.

The visionary "stuff" of the explorationists is not always easy to communicate to senior executives in some corporate headquarters in the South 48. This communications gap clearly influenced which companies ultimately became winners and those who landed on the outside looking in. Happenstance in the offering of leases, and what constituted the "right" lease prices and bids in this God-forsaken operating environment, and even contractual rig commitments, were also determining factors separating winners from losers.

Intracompany cooperation was also a key ingredient in creating a winning team. Exploration people do not work in a vacuum. Multiple disciplines are involved: landsmen, field logistical and drilling people, engineers and others. Exploration cannot go forward without these others to put into action the testing of exploratory ideas.

The Prudhoe Bay discovery was a team effort in the broadest sense of the term. In an interview years after the discovery, H.C. Jamison said, "it [the discovery] is a measure of your capabilities as a professional group. And I stress the word group, because, boy, this thing was a total team situation."

Contributions ranged from minor to major. At important stages, approvals were critical for the project to advance to the next step. The Anchorage explorationists provided good reasons for drilling Prudhoe Bay. They, in turn, recommended it to the middle management of the Dallas exploration staff. It was their responsibility to winnow all projects and recommend the best to executive management. Their stamp of approval was vital and was given. For the Prudhoe exploration well, the top executives had to be convinced to allocate the capital to proceed. They did.

Prudhoe Bay would not have been drilled without the involvement of any one of these three groups. What is not so obvious is the importance of discrete individuals. It is plausible that had some seemingly minor contribution not been made, the drilling project may have faltered.

The only source of geologic structural knowledge in the Colville and Prudhoe Bay areas was the seismograph. This is where the geophysicist part of the explorationist team shines. In oil-field vernacular they are called doodlebuggers. There is no derision implied.

Historically, geophysicists and their families were a hearty lot because they were the advance guard of oil exploration. They moved often and many times to out-of-the-way places. Frequently they were not in a place long enough to establish friendships. They were not patsies, but the Arctic Coast was like the end of the world to them. They then literally lived on the job in buildings mounted on sleds called wanigans, which had been adapted to housing, eating, office, and repair facilities. Crawler tractors pulled these mobile camps, keeping them close to the fieldwork.

Many of the geophysical supervisors during the early years of North Slope exploration in the early 1960s did not have college degrees in geophysics as there were few, if any, colleges that taught exploration geophysics. Because there were physics and electrical concepts involved, these early supervisors had science backgrounds in mathematics, physics and electrical engineering. Many of them had worked directly for some of the pioneers in exploration geophysics.

As the oil companies started to accumulate seismic data in the early 1960s, the huge Colville and Prudhoe Bay structures began to appear on the maps of various companies. These structures were unreal in size. Had they been revealed at the surface they would have been much more imposing than the quintessential sheepherder anticlines in Wyoming.

Paul Lyons, a geophysicist with Sinclair, is given the credit for being the first person to make a map of the Prudhoe Bay structure from geophysical information, which he did in 1963. Several companies, including Atlantic and Richfield, had people making maps of Prudhoe Bay not too long after.

Rudy Berlin was then working for Richfield, but had first mapped the Prudhoe Bay structure when he was working for Western Geophysical Company, a geophysical contractor at the time under contract to Sinclair and BP. He had seen the Prudhoe structures even before the clients BP and Sinclair. Dick Crick, the first Atlantic employee working in Alaska, remembers.

> The first time I heard about the Prudhoe Bay prospect was the summer of 1965 when we had a "get acquainted happy hour" at the Petroleum Club with our Richfield counterparts. I remember that Rudy Berlin mentioned that he hoped the merger of the two companies would not cause them to move off the Slope without drilling a well to test the largest structural closure he had ever mapped. If anyone should be given credit for trying to sell the Prudhoe prospect, in my opinion it should be Rudy Berlin.

At Prudhoe Bay the seismic data looked like a dream come true. In fact to some it looked too good to be believable. Also, the geophysicists did not have precise velocity data to apply to the problem. There was permafrost from the surface to a depth of almost 2,000 feet, and there was uncertainty as to how to deal with this variable, which affected seismic velocity. Permafrost is simply permanently frozen earth. It complicates interpretations if the character and thickness of the permafrost is unknown. It affects the velocity of seismic wave travel through the earth. The geophysicists later learned how this permafrost change from land to off shore influenced the seismic interpretation.

Louis Davis, general manager of North American Producing, was pragmatic and always cut to the heart of issues. At one of our conferences before Prudhoe Bay was approved for drilling, Rudy Berlin presented the seismic picture for the prospect in his inimitable impassioned way. At the end, for additional emphasis, he very forcefully said, "If this was my oil company, I would drill this prospect." Louis Davis turned to Lee Wilson, drilling and production manager for Alaska, and whispered, "I thought this *was* his oil company!"

I am guessing Louis's remark was serious because he wanted team players and Rudy's innocent remark seemed to put him apart, but he was pitching for Prudhoe to be tested like the rest of us were. ARCO had reached a moment of truth. In the district we thought we should drill the Prudhoe structure, and that is what we recommended. It was that or pull out of the North Slope, so we took the positive approach and moved in that direction, as you will see.

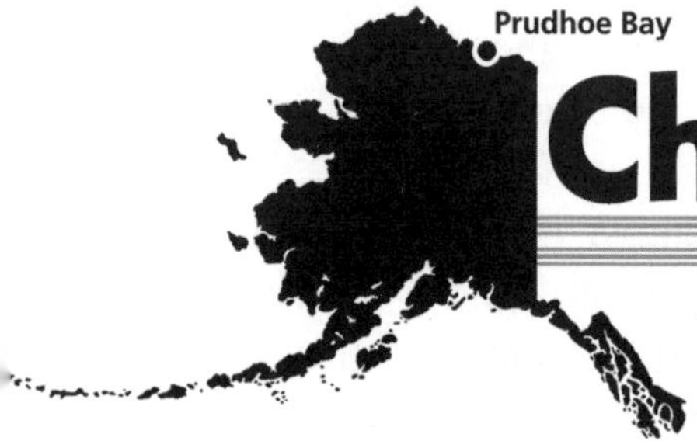

Chapter 12

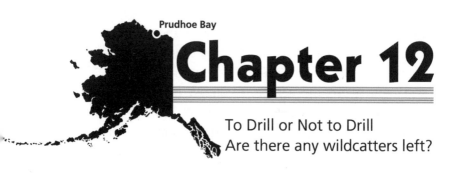

To Drill or Not to Drill
Are there any wildcatters left?

Union Oil of California (Unocal) plugged its *Kookpuk* No.1 as a dry hole in March 1967. The scorecard now read BP: 7; Sinclair: 3; Colorado Oil and Gas: 2; ARCO: 1; Unocal: 1 — all of them failures, dry holes! In four short years beginning in 1963 the oil industry had drilled fourteen dusters in the state of Alaska. The company operators usually assign a chance factor to finding oil before a well is drilled. This is based on their knowledge of the local geology, mainly the quality of the geologic structure to trap the oil, the expectation of the "right" rocks (that is with good porosity and permeability) and the drilling history in the general area. No chance factor is "typical." It can vary widely within an area, between areas, and between people. Only the companies knew what they believed the chance for finding oil was for each of the fourteen dry holes. Suppose that the average chance factor for each of the fourteen wells was 20 percent, or for every five wells drilled one would be a discovery. Fourteen exploratory tests each with a 20 percent chance factor should have yielded two oil discoveries, but there were none. These results came after decades of geological fieldwork by the USGS and petroleum geologists, the drilling program by the Navy in NPR4, geophysical prospecting by the Navy and four more years of seismic exploration by oil companies.

It is axiomatic that companies drill what they believe are their best prospects first. This was the situation facing ARCO's Anchorage exploration staff in late 1966.

If you were the senior executive in charge of finding and producing oil for ARCO, and were asked for $1.6 million — ARCO's share of the cost — to drill another well on the Arctic Slope of Alaska, how easy do you think it would be to give your approval? Remember that the last of those fourteen wells was ARCO's *Susie Federal* No.1. *Susie* had been a Richfield prospect in which the Atlantic top management had no emotional involvement. If they had, as managers in the newly formed ARCO the disappointing baggage of *Susie* easily could have stimulated bias against drilling Prudhoe Bay. Mr. Rollin Eckis (former Richfield President) at this time was a vice president in ARCO, and not in the direct line of authority for this decision, but nevertheless very influential. He still exuded confidence and remained upbeat.

> Don't lose your enthusiasm for this great basin; the area hasn't been scratched...Don't forget we still have Prudhoe Bay.

It sounds like Wallace Pratt, the early dean of all oil finders, mentioned in the Humble Oil and Sinclair chapters. To paraphrase him, don't let too many failures muddle your enthusiasm to keep searching for oil.

The foregoing comments are fine for the lay reader, but for the benefit of exploration-trained readers it is not quite that straightforward. When industry took up the torch after the NPR4 exploration program they chose to look for oil in the foothills of the Brooks Mountain Range, which had easily identified geologic anticlines, and could be determined by surface geologic methods. One such feature was found by NPR4 exploration, and produced oil at Umiat. Eleven of the failed tests were drilled by oil industry members on these prospects. These anticlines are generally believed to be superficial in the surface and near surface rocks, and are not folded in the lower older rocks. Geologically they have a fancy French name, *decollement*. The most effective way to explore the Arctic Coastal Plain on the other hand was by reflection seismograph because there are no rock outcrops to survey for structural strikes and dips. As a consequence, only the ARCO *Susie* and the BP Colville wells were drilled on seismic evidence before the Prudhoe Bay discovery. That does not change the statistical observations made in the paragraph above, but the foothills oil tests and the Coastal Plain tests were two different geologic provinces.

We all know approval was given, and in retrospect even those close to the project probably tend to believe it was an easy decision, but it was not a slam-dunk. Two obstacles loomed. One was the gross cost of the well, $3.2 million; not much these days, but big bucks in 1967. The other was that the ARCO Anchorage staff had to convince not only their Dallas managers but also the top Humble Oil hierarchy.

But back in late 1966, all of us on the Anchorage exploration staff knew it would be a hard sell to senior executives in Dallas. The staff there was composed of Louis F. Davis, vice president for all North American exploration; B.J. Lancaster who replaced Mr. Davis when he was promoted to executive vice president in Los Angeles; J.L. (Jim) Wilson, vice president; Wm. Albright, manager of land acquisition; Julius Babisak, exploration manager; J.A. Savage, chief geologist; John Thomas, chief geophysicist; plus many assistants to each of them. They were all very astute and the leader, Mr. Davis, demanded answers, and woe betide anyone who tried to flimflam him. My colleagues and I knew we had to review every bit of information and be as convincing as we possibly could, or in pipe-organ jargon, pull out all the stops.

As noted, explorationists must possess great optimism; on the other hand, the Dallas people holding the authority for final approval had to weigh all of the projects presented to them and then allocate funds company-wide to those projects having the best possible rewards and the lowest risks. In arriving at a decision, the vice presidents and chiefs would use their experience to probe deeply into the science and optimism of the explorationists. We counted on it as being a hard sell. Their questions would come in different ways and from different points of view. They would test our various theories for weaknesses, and even our resolute enthusiasm. They would challenge the level of commitment and intrinsic belief in our project's potential for success. The pressure would be palpable.

It didn't help our cause that the very large, classical potential oil structure just west of the Prudhoe structure (Figure 14), the Colville

Figure 15: A north–south (A–A') cross section approximately through the center of the Prudhoe Bay oil field. The location of this section is shown on both Figures 14 and 21. The faults are labeled near the north end of the section. Note that the Colville formation gets thinner from south to north. The Kingak shale below the Colville Group disappears completely. This thinning means that the Prudhoe Bay structure had formed before the Colville Group was deposited. Looking closely, one may see the positions of the water, oil and gas zones. Notice the vertical scale is eight times the horizontal scale. This is a good example of the distortion of cross sections. These few inches of diagram represent a distance of just over fifty miles.

After Jamison, H.C. et al. Giant Oil and Gas Fields of the Decade 1968–1978 *AAPG ©1980 reprinted by permission of the AAPG whose permission is required for further use.*

Figure 16: An east–west (B–B') cross section through the Colville and Prudhoe Bay structures. The cross section extends from the Union Kookpuk *No.1 on the west to the ARCO* Prudhoe Bay State *No.1 on the east. The gentle arching just east of the Union* Kookpuk *is the Colville structure we have mentioned so often. The Colville Group thins from west to east as Figure 5 shows it thinning from south to north. The water, oil and gas zones appear on this figure too.*

After Jamison, H.C. et al. Giant Oil and Gas Fields of the Decade 1968–1978. *AAPG ©1980 reprinted by permission of the AAPG whose permission is required for further use.*

Figure 15: North–South cross section

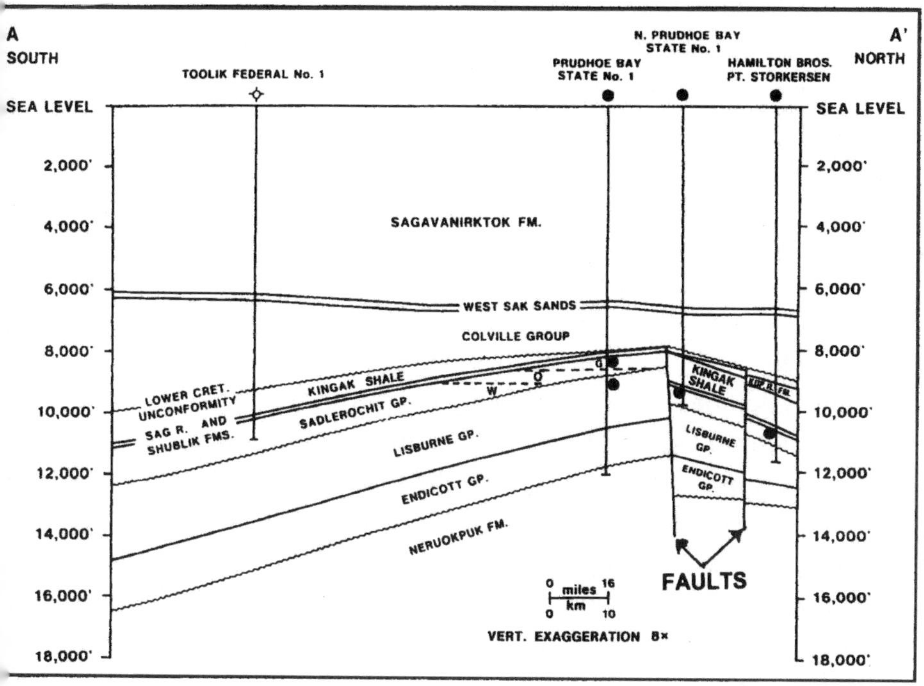

Figure 16: East–West cross section

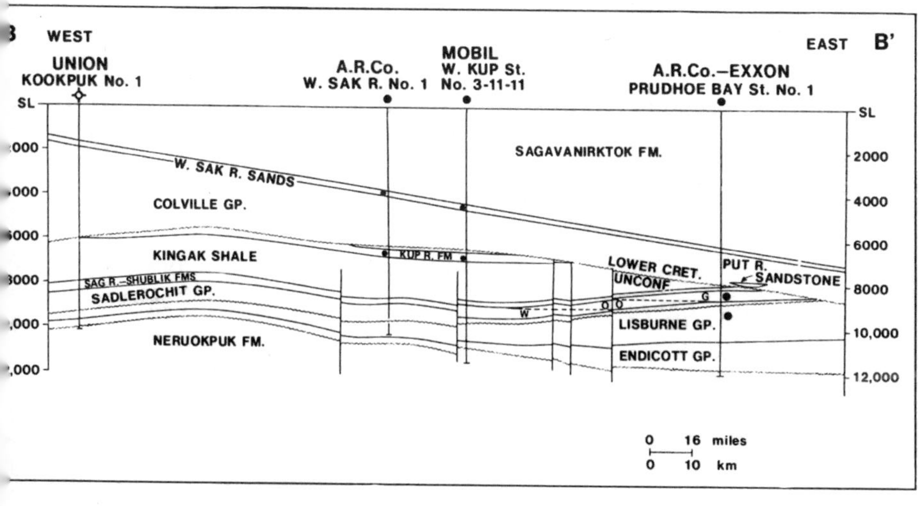

High (Figure 12), was more than a great disappointment; it was a total failure. British Petroleum, a frontrunner in Arctic Alaska exploration closed its Anchorage office as a result of the Colville failure. The decision of the Dallas managers could have been foreordained. "So why are you coming to us asking for money? Are we missing something? Why should we take the risk when a company like BP is packing it in? This has all the earmarks of another monument to science!" These all would have been understandably logical, anticipated questions. Fortunately for us, the Dallas folk were open minded.

The term "monument to science," usually said with a sneer, refers pejoratively to a dry hole with no redeeming virtues. Many dry holes have, however, added to the science and art of oil exploration, having provided clues to future discoveries. The previously mentioned fourteen dry holes added geologic stratigraphic information far from the surface outcrops, which had been studied throughout the years. The tests also yielded important seismic velocity data to aid in understanding the seismic maps. This information remained confidential among the operators. In any event, none of the wells gushed forth the silver bullet pointing to a new theory, much less a new discovery.

Our case was not ironclad. Seismic data was expensive to obtain, so there were never as many survey lines as we would have liked. More data would have given us more confidence. The disappearance or curtailment of seismic events almost certainly was caused by a truncation of the rocks underground. (Figures 15 & 16)

There was also the likelihood of faulting. Faulting occurs when dynamics in the earth cause so much force the rocks break and slide in different directions on each side of the faulting zone. Which would it be here, faulting, truncation or both? The dip interpretation of the seismic, which would prove trapping of the oil in the northwest, southwest and southeast, was rock solid. Dip closure means that from the crest of the subsurface anticline (arch in the rocks), the surface of those rocks sloped away from the crest to the southeast, southwest and northwest. It is within the confines of this slope, or dip, of the anticline that the oil is found. (Figures 14 & 15) Seismic could reveal only so much. Changes occur in rocks hidden under the surface, which are undetectable by seismic and this could greatly alter the interpretation. What were the rock characteristics really like in this questionable area? Figure 14 is the seismic interpretation before the discovery well was drilled. Figure 21 is what it looked like ten years later and was based on additional seismic and data from the well that had been drilled.

There were differences between the Prudhoe subsurface geologic structure and the dry Colville High structure, in spite of the fact that they were near each other. The Colville structure is a fairly classic, uncomplicated anticline, almost domelike. For the layman, the Prudhoe structure might be likened to half or three-fourths of an anticline. The structure is about twenty-five miles long and roughly one-third as wide. Only a hole in the ground would tell.

Part of every exploration drilling project recommendation required a justification letter. As a condition of the sale of ARCO to BP, Phillips petroleum acquired the ARCO oil production and all the Prudhoe Bay files. Dick Crick, an ARCO senior geologist during the Prudhoe days, was given permission by Phillips to search the files in 2000 hoping to find the discovery well justification letter. Sadly, it was not found. The district geological staff had the responsibility of preparing the drilling recommendation letter. At that time I was district geologist. This letter was a narrative of the salient parameters necessary for approval. Covered were the objective oil formation, its thickness and depth, geologic structure details from seismic maps showing the expected area and shape of the structure, a prognosis of the geologic character of rock formations and specific key formations from the surface to the expected total depth, and the location of suspected zones where drilling might be hazardous.

The chance for success and expected reserves were the most important ingredients. I estimated the chance factor at 35 percent. H.C. Jamison confirmed that figure in later years in correspondence. In other words, about one out of three similar situations should yield a discovery. This is a very high chance factor. From another perspective, if Prudhoe were combined with the fourteen pervious dry holes, the discovery rate for the whole Arctic slope area would be one in fifteen, or 6.7 percent. The 35 percent figure could easily be challenged because it might be believed to be too high, but consider the following: some of the other dry holes in the area had some indications of oil, and the Prudhoe structure was spectacular compared to all the others except the Colville structure. Based on this, a one-for-three discovery ratio does not seem out of line. Success has its own confirmation.

Being a two-handed geologist there is an "on the other hand" way to look at chance factor for Prudhoe Bay. One might say that because the Arctic Coastal Plain was in a different geologic province for all but two of the failed tests, a third test for a different structure in the Coastal Plain area might be expected to be one for three, or rounded to 35 percent chance to find oil.

When we presented the recommendation to the Dallas exploration staff I don't remember there being any disagreement about the chance ratio.

The percent chance factor notwithstanding, nothing like this had been tested on the Arctic Slope. Prior to drilling, the main objective was the Mississippian Lisburne limestone formation, which was estimated to be at a depth of approximately 9,000 feet at Prudhoe Bay, and I estimated reserves at two billion barrels of oil. Mississippian is a geologic period, which in the geologic grand scheme of things occurred between 325 and 360 millions of years ago. Frank Schrader named the Lisburne limestone while on his 1901 expedition. He named it Lisburne because he saw a thick column of limestone at Cape Lisburne on the Northwest Alaska Coast.

How did we estimate that the Lisburne limestone might contain two billion barrels of oil? Primary oil recovery ranges from carbonate rock (limestone and dolomite) in producing formations are well known from reservoir studies of other oil-producing areas. It is customary to pick some reasonable recovery within known ranges and apply that factor to an unproved area. The unit of measure is barrels of recoverable oil per acre-foot. Visualize a square acre of rock that is one-foot thick and, ideally, looks like a sponge (rock) full of oil. Another way to think of it is as a tank of oil one square acre in size and one foot thick. This volume is 7,758 barrels of oil. Assume that the one-foot-thick tank is instead, one foot of rock with 10 percent porosity. Oil in place then would be the product of 7,758 x 10% = 775 barrels per acre. Many factors affect the amount of oil recoverable. Recovery of oil in place is almost never greater than 50 percent on the high end, and oil can be produced at lower percentages down to a point determined by economic factors. Pay thickness, the rock thickness containing oil, also has to be estimated, based on the experience in known accumulations. At Prudhoe it was based on the known thickness of the Lisburne from the closest surface geologic work and the seismographic interpretation.

The one thing we had firm information about was the maximum closure of the Prudhoe structure (Figure 14). Closure can be likened to the topographic contours around a hill. Imagine that the lowest continual line of equal elevation around the hill is 1,000 feet, and that the hill is 2,000 feet high. That would be equal to 1,000 feet of closure if it were in a subsurface geologic structure. The Prudhoe closure was not certain information, but reliable, based on the quality of the seismograph data.

The horizontal area of the structure from the lowest closing contour (similar to the hill example) to the crest of the structure on a seismograph map is easily compiled using the map scale. This feature, full of oil, is described as being full to the spill point.

Let's digress here to examine a very elementary example of an oil field (Figure 17). The plan, or map view lowest in the diagram, and the cross section above it represents a classically simple oil accumulation. The lines that look like an elongated bull's-eye or target are contours or lines of equal elevation. I have left them unlabeled as it will be sufficient for this explanation just to know relatively what is lowest and highest. At some point inside the smallest circle on the map view will be the highest or top point of the anticlinal structure. Somewhere below the third closed contour will be the lowest closing point. Depending on what is known about a real structure, this distance can vary. In a large structure, which Prudhoe Bay turned out to be, it could be hundreds of feet. In a small structure it might only be tens of feet. Here it is between the third contour and the fourth contour, so lets just arbitrarily think the lowest closing contour is half way between three and four and call it 3.5.

In a classically simple situation three things must occur to form

Figure 17.1: Oil field north of Lander, Wyoming, easily seen west of state highway 789. See also photo no. 50 in color section.
Photo: John Sweet and Jack Nason

an oil field. First, there must be a source of organically rich rock, usually shale, because that is the kind of quiet depositional environment where organisms live. As oil is formed it rises because there is also water in the rock; the oil will keep rising to get on top of the water, until something stops it. (We will return to this subject on page 191.) The second thing needed to form an oil field is a reservoir of not only porous but also permeable rock in which the oil can collect and migrate, until it is stopped by the third necessity, and that is a trap in which oil accumulates and cannot go farther.

In an effort to help lay readers better understand some of the ins and outs of oil accumulations in a real-world setting, Figure 17.1 shows an oil field north of Lander, Wyoming, which depicts some of the things outlined in Figure 17. As this oil field only covers 850 acres, about the equivalent of a square mile and a quarter, all of it is visible in the picture. It is nearly the shape of what was thought of in the early days as a "dome-al" accumulation; think of an inverted coffee cup. (Readers are also directed to the color version of this image in the color photo section of the book.) A small oil derrick is visible near the right center of the picture. Upwards, and a little to the left, is the faintly red outcrop of sandstone trending on upward toward the center of the scene on the horizon. The red lines drawn on the picture are the approximate tops of formation beds that arched over the structure before erosion removed them. The stream just below left center flows in the valley to the right across the anticline.

When geologists work in the office (or, in the early days, surveyed on the ground) they formulate their ideas on maps, using cross sections to convey the third dimension. While the maps are a convenient way to represent the geologic structures in nature, the scale of the maps also accentuates the data in a way that highlights what is important enough to pursue.

With that in mind, the anticline near Lander was self-evident when it was seen, explored, and oil discovered in 1912. The seismograph was not in use at that time, but structure of that size, if detected by the seismic explorers, may not have appeared to be of sufficient size and reserve capacity to justify drilling an exploratory well.

This oil field is visually impressive to see on the surface. By comparison, Prudhoe Bay covers about 200 square miles; but Prudhoe Bay does not have a surface manifestation. Had Prudhoe Bay been explored with too small a grid, only an unimpressive part of it may have been seen and remained unrecognized as a part of very large, intriguing geologic structure.

During its life, the Lander field has produced 16 million barrels of oil. Lots of oil, one might think. It is and, at 2007 crude oil prices, it must be profitable. Looking at the bigger picture, in August 2006 the United States used 22.6 million barrels of oil every day; however, national production was only 8.9 millions barrels of oil a day. Nine percent of that came from Alaska, and most of that from the Prudhoe Bay area. The remaining 91 percent came from hundreds, or perhaps thousands, of little fields like Lander and all the fields between them and Prudhoe Bay. These statistics alone make a case for the nation doing all it should to keep

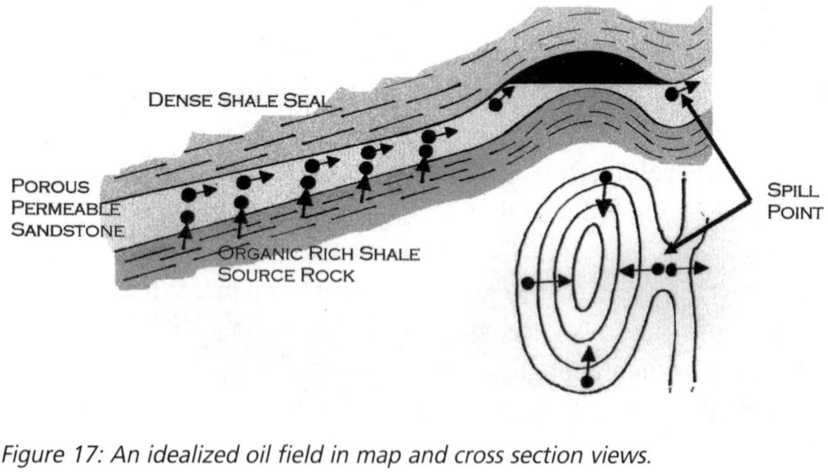

Figure 17: An idealized oil field in map and cross section views.

from falling farther behind in the need for liquid petroleum. We (collectively) are whistling by the graveyard.

Without the Prudhoe Bay contribution over the last nearly thirty-year period (to June 2007), it is hard to imagine how the rest of the oil fields in the country would have been able to produce the equivalent of the 14 billion barrels that have been pumped through the Alaska Pipe Line, or to even comprehend how much less they would be capable of producing now, had that been the case. Nationally the United States can never be oil self-sufficient, but we should be doing what we can to help by exploring in the Arctic National Wildlife Range and in various places on the outer continental shelves.

All the foregoing is just another reminder what an unbelievable stroke of fortune the Prudhoe Bay discovery has been. No grouping of superlatives can capture its significance.

Now to return to the "until something stops it" topic from page 189. In the cross section of Figure 17, the arrows representing the moving oil will rise first vertically until something stops the motion. In the cross section this is the dense shale seal. That seal stops the vertical movement, but the oil continues to be buoyed upward by the presence of water, except now it moves laterally (to the right in the cross section) under the shale seal. Now think of the elongated contours of the map view as an elongated upside-down cup or, better, an oval gravy bowl without a spout. The oil stops and collects in the dome of rock under the surface of the land and is trapped there. In our example it will accumulate until the trap is filled to the last closing contour, which we set at the estimated line of 3.5. If oil continues to be generated, it will begin to leak at the spill point and will continue its course on up and dip beyond the spill point toward the right edge of the map view and the cross section.

Putting it all together then, the area in acres of closure, multiplied by the average thickness of the Lisburne, multiplied by the recovery of oil per acre-foot, equals the total estimated reserve. Maps similar to Figure 14 would provide some of the source information needed to estimate the potential reserves at two billion barrels. No one in the company had ever been associated with a field even a small fraction as large as two billion barrels. There was never any feedback about that estimate, but I have always doubted that anyone took that number seriously simply because it was so far out of our collective experience, and so long since such large reserves had been discovered in the United States. Can you imagine what they would have thought if we had said 10 billion barrels, which it proved to be? Someone would have called for straightjackets.

Julius Babisak, the exploration manager in Dallas, was also preparing for the event. ARCO had planners and statisticians who wanted to quantify exploration so it could be massaged statistically and quantified in a business plan. Julius was very astute and would never let himself be boxed in by their requests. He was a quick study and would find unique ways for handling planning requests. Here is how he handled this one.

> The planning people had come to me and said, "Look, if you push this project you're going to have to guarantee that those wells make 500 barrels each." I said, "You've got it, it's 500 barrels." I didn't care what it was, because it was either nothing or something. I didn't know there were reservoir rocks there. I didn't know there was anything there. For me to say that 500 barrels was okay was not done on the basis that that was the case, I just knew that I didn't want vague numbers to jeopardize the project.

At first glance you might wonder about this kind of reasoning, but it served the planners' purposes. Some of the wells went on to produce at the rate of 23,000 barrels per day, forty-six times what the planners said was necessary. Do you think they would have believed Julius if he said 23,000 barrels of oil per well per day? Almost certainly not. As for the 500-barrel-per-day bone Julius threw them, in retrospect that would not have come close to being commercial if that were all that had been found. So much for planning and statistics. This is another of the many ironies.

Jim Savage, chief geologist, put it a little differently. "Geologists are not comfortable with numbers, we're not comfortable quantifying things. While we recognize the need to be economic we know we can't really quantify nature...we realize numbers that we give fall in a wide range...they are just best guesses." I always suspected that everyone involved believed that if the test well hit oil, it would have to be very large, so there was no need to quibble on a specific number.

All of the above set the stage for the most important exploration conference ARCO was to ever have. This was the "make or break" effort to convince ARCO management, and our partner Humble (Exxon), that we needed to test the Prudhoe structure. The dates were December 2–3, 1966.

On December 1, a group of us flew to Dallas. We had permission to fly first class because there was a five-hour time zone change and about seven hours in the air, plus time between planes. Occasionally some would take the red eye, which meant flying all night and then attending meetings when arriving in Dallas in the morning. Other times we would overnight in Seattle. This was too important a meeting for that, so we left earlier than usual to dampen the jetlag. The Anchorage contingent included: H.C. Jamison, Alaska district manager; C.H. Selman, district geophysicist; Roland F. Champion, district landman; Leland B. Wilson, district drilling supervisor; Marv Mangus, senior geologist; and me, district geologist, all prepared to convince our senior Dallas management to drill. We did this on December 2, and the next day we made our presentation to the Humble (Exxon) people headed by Tomas Barrow, their top exploration vice president.

All the pre-conference preparation paid off. We had the answers to all the questions that were answerable. For those to which there were no answers, we gave our best estimates about what we might reasonably expect to occur. There were a couple of outstanding questions about leasing that also tilted in favor of drilling. There had been unleased state acreage tracts near the proposed test well location. H.C. Jamison appealed to Governor Walter J. Hickel, asking the state to offer the tracts for lease. A test well of the Prudhoe structure would give a fair evaluation of the oil potential of these tracts, and we believed ARCO or someone needed to have leases on those tracts if we were going to take the risk of drilling. Harry and the land people cleared that up.

> The acreage was put up by the State; we did acquire it. It would have been a very, very awkward situation then to have backed out of that sort of virtual [implied promise to drill a well] commitment.

Another commitment similarly figured into the equation. In *Prudhoe Bay, Discovery to Recovery* by Gene Rutledge, E.M. "Mo" Benson is quoted.

> One of our obligations to the drilling company was to return the rig to Fairbanks if we didn't drill a second well. According to Brad [T.F. Bradshaw, ARCO president] he used that argument to support drilling of the *Prudhoe Bay State* No.1, since, in effect, he considered it a free well as we would have had to spend the same money to fly the rig out.

Mr. Benson at this time was a vice president in charge of Alaskan activities.

In final analysis, the Prudhoe structure was just too large and looked too good not to warrant a test. Everyone involved likely recognized this simple rationale. The consequence of not drilling was perhaps a more significant factor in the decision to drill than many realize. There was no way to get more definitive information. No other company would be doing any kind of work that would yield additional knowledge. Future costs would be greater still, because a rig would again have to be mobilized and brought in from the Cook Inlet area, or Canada, or who knows where. It is very likely that the well would never be drilled if ARCO did not seize this opportunity. There were more reasons to drill than not to drill. An item that was not discussed, and in retrospect had relevance, was the fact that some of the geophysical shot holes encountered gas and shows of oil.

Other details of the meeting escape me, but there were no hang-ups. Lee Wilson seemed to think we did okay. Author Booton Herndon quotes Lee's take on the meeting in *The Great Land.*

> John Sweet came to me with an idea. He said we had one last chance to hit a big field on the Slope. I've heard geologists talk before, and I take their dreams with a grain of salt. I guess I am a pessimist. But all he wanted me to do was to back him up with some facts and figures. I had them and I went along…John made a two-hour presentation to Louis Davis and his staff. He gave it all he had. I told you I was a pessimist, but before he got through I was raring to go.

There was one final hurdle, the corporate executive committee's approval. Louis Davis credited W. Dow Hamm, executive vice president and a board member and in charge of domestic and foreign exploration and production, with handholding the committee. He spoon-fed them, as it were, because they didn't really understand the exploration side of the business. At the same time they realized they needed oil exploration and had to have faith that we knew what we were doing. They were very nervous. Big and risky expenditures had that effect on people, especially the conservative easterners who were used to dealing with more predictable refining and marketing issues. Louis Davis later said:

> But I remember that I was told, if this one is dry, don't come back. This is it. We don't want to hear any more about the North Slope of Alaska.

Imagine what would have happened if somewhere during all this decision making someone had said, "By the way, if we find oil it will cost untold billions to drill development wells and fabricate production facilities and still another eight or ten billion to build a pipeline." Mr. Robert O. Anderson was never quoted in any company writings of which I am aware. Recently, when reading *The Prize* by Daniel Yergin, I was surprised to see this quotation attributed to Mr. Anderson, which he subsequently confirmed.

> It was more a decision not to cancel a well already scheduled than to go ahead.

After obtaining all the necessary in-house and Humble approvals to drill, ARCO asked British Petroleum for a contribution. It was customary when drilling an exploratory well to seek contributions from nearby leaseholders in exchange for giving them the well information on a timely basis. These contributions can be either monetary or acreage. BP indicated they would give a one-half interest (called a farm-out) in a four-section (2,560 acres) that was directly southwest of the ARCO drilling location. BP had paid $46.60 dollars per acre, or total bonus for this lease block of $118,048. Richfield paid a lease bonus price of $93.78 per acre for the lease on which the discovery well was drilled, almost double the price BP paid. These differences in what the two companies paid for adjoining leases does not represent any particular difference in knowledge. After thc field was developed, the BP lease for which they had bid less was much more valuable than the lease for which Richfield had paid the highest lease bonus.

Our executives flatly rejected this offer (half interest in the 2,560-acre adjoining lease) as being much too small. The BP offer was all upside potential for ARCO because a discovery would have added to the ARCO reserve position. Here it would have been tremendous, because the BP leases were more favorably located on the structure than ARCO's and thus had much more oil under them. Some of those four-section blocks had a full oil column (this means the total thickness of the formation held oil), and the average block recovery was 250-million barrels or better. In the South 48 this is equal to a couple of major oil fields. At a later price of $20 per barrel, that is $5 billion. What a huge irony, they believed it to be too small.

The previous observation is obviously hindsight, but weighing the same alternatives was an option before the fact, and those responsible apparently never looked at the upside potential. ARCO had cut off its nose to spite its face. It got nothing from BP. And BP got nothing from ARCO, but when all was said and done BP came out the winner because ARCO had proved the BP leases. In due course, BP got the well information just for reproduction costs.

The decision to drill the *Prudhoe Bay State of Alaska* No.1 now had all the stamps of approval. Richfield had started the *Susie* prospect well just shortly after the Atlantic–Richfield merger. That test was a dry hole, so after the abandonment of *Susie* on January 9, 1967, the contractor and the drilling engineers began moving the rig to the Prudhoe Bay drilling location.

Chapter 13

Arctic Roulette: Drill the Prudhoe Structure

The approval hurdles were conquered, drilling would soon begin, and the Prudhoe Bay structure would be tested. But before a rig is moved and drilling begins, the drilling contractor must be told where this bold adventure is to be located. This process is called staking the location because in the South 48 a surveyor is sent to an area to survey the land and locate the precise drilling spot. When that place is found, he will drive a surveyor's stake into the ground and identify it with the well name and appropriate surveying data. Nearly always, the stake will be adorned with orange or yellow flagging, loudly proclaiming itself.

In northern Alaska, where land surveying was in its early evolutionary stages, staking the location was done differently. The task was delegated to Marvin Mangus and Wray Walker. Marvin's love affair with Arctic Alaska dated back to 1947 with the USGS in NPR4. Wray, an ARCO geophysical crew supervisor, had spent much time in the Prudhoe Bay area. He was one of those individuals who by his performance had stimulated trust in his ability to get the job done. He was an experienced surveyor, and his specific duty was to guarantee the validity of the contract seismic crew's work. The seismic interpretation determined the best location to drill, so it was Wray Walker's responsibility to make certain the stake was placed

Top photo: Governor Hickel's visit to Prudhoe Bay State No. 1 May 2, 1967. Standing left to right: Bill Eiting, Humble unidentified truck driver; Bob Fawcett, pilot; J.B. Coffman, Humble western exploration manager; Lael Morgan, reporter Fairbanks News-Miner; Jim Magoffin, president Interior Airways; Nibert Johnson, Humble western land manager; Neal Bergt, pilot; Roland Champion, ARCO district landman; H.C. "Harry" Jamison, ARCO district manager; Walter J. Hickel, governor of Alaska; Bob Walker, Humble district landman; John M. Sweet, ARCO district explorationist.
In the doorway top to bottom: Roscoe Bell, director State Division of Lands; Alaska trooper; governor's escort Phil Holdsworth, Commissioner of Alaska Department of Natural Resources.

Middle photo: Left to Right John M. Sweet, ARCO district explorationist, Governor Walter J. Hickel, Roland F. Champion, district landman, and H.C. "Harry" Jamison, district manager.

Bottom photo: The Prudhoe baker greets Governor Hickel with a special cake.

in the physical location that best matched that on the seismic map. The importance of this responsibility is self-evident. In retrospect, however, a dart dropped from a plane flying at 2,500 feet in the general area would have accomplished the same purpose. The future oil field turned out to be that large.

When Wray and Marv pounded the all-important stake into the ground, the contractor was finally allowed to move the drilling rig over the location. Then the massive substructure that supported the rig was carefully maneuvered over the stake.

The ARCO–Humble *Prudhoe Bay State* No.1 was "spudded" on April 22, 1967, at 7:30 in the morning. Spud, or spudded, is oil-field jargon for to start or started. In the very early days of drilling, wells were started with a steel hand bar called a spud. The bar is no longer used, but the long-standing term is entrenched. On May 2, the drilling crew reached a depth of 579 feet and then abruptly stopped while still drilling in permafrost. The warm mud was thawing the permafrost, causing earth and rocks to fall around the drilling bit. Progress came to a halt.

That same day a delegation of ARCO and Humble people took Governor Walter J. Hickel and two staff members, Phil Holdsworth, the commissioner of Natural Resources, and Roscoe E. Bell, the manager of the Division of Lands, on a goodwill trip to the drill site. Bob Walker, Alaska landman; Nibert Johnson, western operations land manager; J.B. Coffman, western operations exploration manager; and Bill Eiting represented Humble Oil and Refining (Exxon). H.C. Jamison, Alaska district manager; Roland F. Champion, Alaska district landman; and John M. Sweet, Alaska district explorationist, represented ARCO. Lael Morgan, of the *Fairbanks News-Miner,* reported on the trip.

While others' recollections may differ, I vividly recall the governor declaring, "Gentlemen, we are standing on top of one of the greatest oil fields in North America." In later years, Governor Hickel was quoted as saying forty billion barrels, which I missed because it was said out of my earshot on the plane returning to Fairbanks. He said it in conversation with Harry Jamison and others. Here is the governor's recollection in a 1986 letter to Mr. Gene Rutledge.

> I was positive there was oil beneath the surface. For me, Prudhoe Bay wasn't a gamble; it was an absolute. In fact, I felt so strongly about the presence of oil that I predicted that there were 40 billion barrels of reserves and that we could recover half of those.

The day was sunny, with temperatures ranging between fifty and sixty degrees. Most of us were comfortable in sport coats, sweaters or wool shirts, even at the rig site. The landing strip (Figure 18) was on two frozen lakes with a bladed tundra strip connecting them. The group traveled in a DC-3. H.C. Jamison remembers there was water standing on the lakes when we landed. By the time we left in mid-afternoon, several more inches of water stood on top of the ice.

Break-up is the name given to the period when the thawing of snow and ice begins in earnest, or when ice in major streams begins to move downstream. Break-up on land usually occurs at indefinite times and for indefinite periods, but this day and other warm days like it had the appearance of an early break-up. The drilling people were uneasy about the nice weather. They envisioned an unstable drilling situation on their hands. If break-up really was in progress, the airstrip on the lake would disappear, and with it reliable transportation. In the event of serious trouble, no certain method of transporting supplies and equipment to the rig would be available. The following day, May 3, 1967, the Anchorage drilling department, supervised by Lee Wilson, suspended drilling.

After the ground and lakes had frozen in the fall, drilling resumed on November 18, 1967. The drillers analyzed the previous spring's difficulties and made some changes in technique in order to avoid similar problems. On November 25 they had set and cemented casing at 2,035 feet. Casing, another name for pipe, encases the hole to make for a more stable drilling situation than would exist if one were drilling in an unprotected rock section. At the depth of 2,035 feet, it was referred to as surface casing because it protected the near-surface part of the drill hole.

Before reaching that depth, and while drilling at about 1,000 feet, large pieces of wood circulated out of the hole. A thirty-inch piece was sent to Julius Babisak, exploration manager in Dallas. He had some of it cut and mounted in Lucite for conversation mementos. He also sent a portion for analysis. The sample was dated at 42,000 years and identified as tamarack, the same species existing today. Talk about global cooling; there are no longer

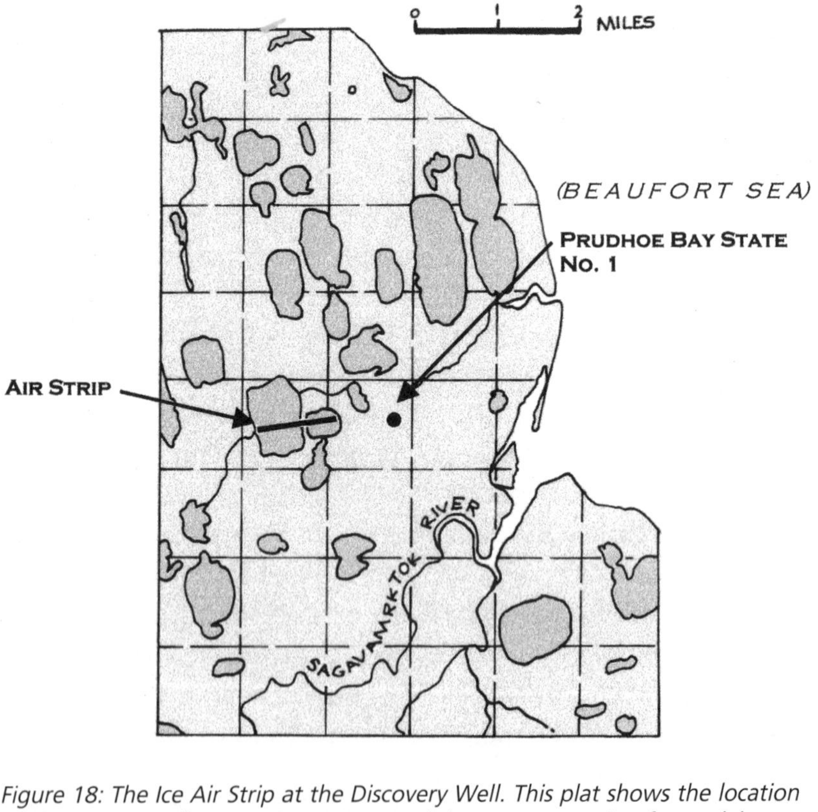

Figure 18: The Ice Air Strip at the Discovery Well. This plat shows the location of the ARCO Prudhoe Bay State *No.1 and the two connecting frozen lakes used for the one-mile airstrip. The area of this plat contains over 30 lakes. The Arctic Coastal Plain has literally hundreds of lakes. The mosquitoes love it.*

similar forests of tamarack growing north of the Brooks Range. On the other hand, it wasn't growing where it was found, but it would have been ground to splinters if it had been transported too far. It does give a good idea of the rate of sedimentation for that area. One thousand feet is a great thickness, but spread over 42,000 years, the amount is less than three-tenths of an inch per year.

By the time we arrived at this momentous drilling venture our staff was thoroughly familiar with the background geology and geophysics. We had experience working with the tools used for evaluating the oil potential of exploratory wells. One of the most important of these is a drilling mud logging unit or, in the vernacular, a "gas-sniffer." For those unfamiliar with the drilling of oil wells, the mud-logging unit was usually provided and operated by a contract service company. Mud is the common term for drilling fluid. Engineers, trained to examine the rock chips carried from the bottom of the hole by the circulating drilling fluid (mud), manned the units. The most important tool in the logging unit is the gas chromatograph (Figure 19), from which stems the term "gas-sniffer." A chromatograph can detect minute quantities of gas in the drilling mud. When these gas shows have what is believed to be a sufficient combination, intensity and thickness, the geologist will likely ask the drillers to stop and take a test. And while most of what the mud engineers do is a duplication of ARCO geologists' work, the chromatograph is the primary reason for utilizing the units.

On this well the service company was Core Laboratory. D.M. Trainer, lead engineer, and D.L. James, assistant engineer, operated the unit. These contract operators were under the direction of the company well-site geologist Marvin Mangus, ARCO senior geologist William C. Penttila, and staff geologist Robert Anderson. In oil jargon they were "well sitters." The men alternated between home-sitting in Anchorage and well-sitting at Prudhoe Bay, overlapping only when changing places.

Figure 19: Mud Log

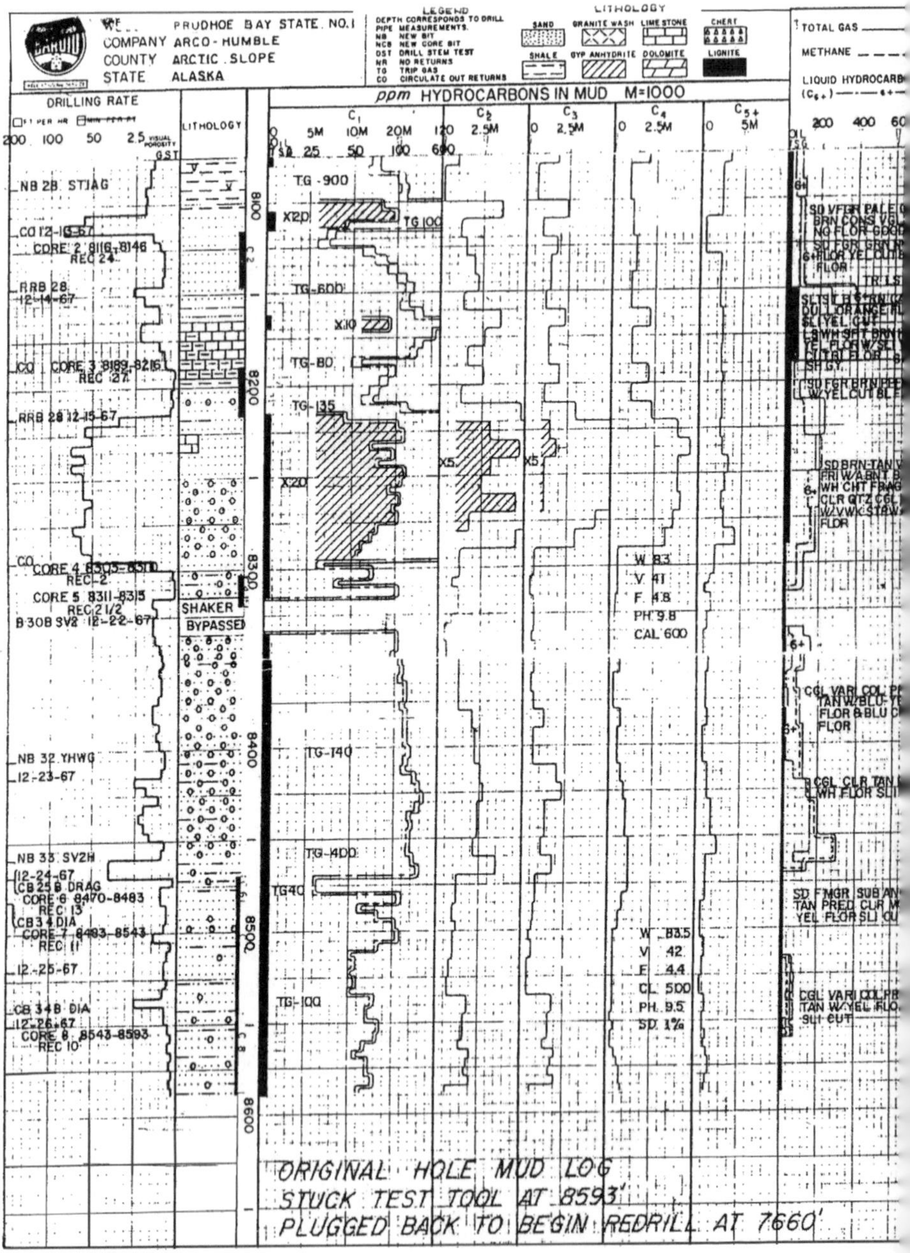

A vigilant well-site geologist constantly watches (Figure 19) the rate of drilling. This is where the proverbial buck stops, or starts in this situation, because he calls the evaluations shots. A change in rate, called a drilling break, is occasionally a good omen, especially if the rate increases. The rate is obvious by recording the feet per minute, and also by watching the drill pipe turning on the rig floor. In very slow drilling, the pipe may move down the hole imperceptibly; with very fast drilling the drill pipe moves down the hole very quickly. Usually this points out that the rock being drilled is more porous. A drilling change can also signify a change in geologic formation. The well-site geologist usually knows the expected sequence of formations from other wells from surface geology, or from the rocks he has already seen in the subject well. When well loggers or engineers see a drilling break, they commonly have standing orders to notify the ARCO geologist, who may decide to have the drillers stop drilling long enough for the cuttings from that depth to be circulated to the surface so they can be evaluated for oil shows. The geologists can determine if there are shows of oil from the appearance of the well cuttings and with the help of a fluorescent

Figure 19: Cutting analysis log of the discovery. This is an example of a mud log, an analysis of the drilling fluid, "mud" in the vernacular. Drilling mud has several functions, it cools and lubricates the drilling bit, it carries the well rock cuttings to the surface, and it helps control high-pressure zones of oil, gas or water from causing a "blow out" or loss of control. The mud log is one of the most important tools for determining the presence of oil and gas. From left to right, the drilling rate records how quickly (to the left) or slowly (to the right) the drilling is progressing. The lithology records the kind of rock being penetrated. The ppm hydrocarbons records the gas analysis of the mud as read by the chromatograph. The C1...C5 curves record different levels of gas concentrations from C1 methane through the heaver gases. At 8,100 and 8,210 the C1 readings were so intense that the operator had to increase the scale twenty times to plot the curves. The final reading on the right is a cutting analysis. A blender pulverized the rock cuttings. Bill Penttila, ARCO well-site geologist, said that in the zones with the best oil shows, so much gas was liberated, the lid would blow off the blender. This resulted in a mess of mud and cuttings covering everything in the mud-logging trailer.

light. Oil causes a yellow-to-gold fluorescence when viewed under a fluorescent light. Carbon tetrachloride is a solvent that will free or release the oil from the rock fragments. This will, in turn, cause a streaming cut, which will show up under the fluorescent light. In a glass sample dish, about a couple of inches in diameter, it might look something like a small yellow or gold stream. It is actually the oil released from the rock floating on water. Occasionally, if there is enough oil and the rock releases it easily in the presence of carbon tetrachloride, one may see very small globules of oil. Ordinarily, gas shows are the first good indications of hydrocarbons (oil and gas) because gas is liberated more easily from rocks than oil is.

Armed with these tools and skills, the mud engineers and the ARCO geologists were ready for the drill cuttings to begin to yield the secrets of the Prudhoe Bay structure. The evidence was not long in coming. On November 29, less than four days after the drillers set (installed) the surface casing, the well began to encounter shows of oil and gas. The shows occurred at 3,300 feet in the shales, siltstones and friable sandstones of the Sagavanirktok formation in the Tertiary. These were manifest by modest readings on the various gas chromatographs and were both interesting and encouraging. Unfortunately, they were not in rocks with sufficient porosity to hold much oil nor with the kind of permeability that would allow oil to flow to the borehole. These shows were good, but not yet enough.

Below 3,300 feet, the drill bit continued to cut through rocks that still did not have redeeming reservoir or permeability qualities. It was encouraging that the shows not only continued but also increased. And while not exciting, as the well approached depths of 4,000 feet then 4,500 feet, the continuing oil shows began to cause a stir and some head scratching. In truth, this substantial thickness of non-reservoir rock, but with good oil shows, was beyond the scope of our collective experience. What was the meaning of this situation?

Five thousand feet and there was no change in the reports from the field geologists Marv Mangus and Bill Penttila. Finally, there was a change; the oil shows improved. From 5,335 to 5,390 feet, geologists at the well began to see a light brown oil stain, some dead oil, free oil content of 2 percent, and gold fluorescence. Although the bit had drilled into older rocks of the Prince Creek formation (the Colville Group of the Upper Cretaceous), the lithologic character of the rocks had not appreciably changed. They were not good reservoir rocks.

Two percent cut might not sound dramatic to the layman, but it is to the expert. Here is the picture: carbon tetrachloride is a solvent geologists use to coax oil shows out of the drill cuttings. The cuttings are very small, generally pencil-lead size, and the grains making up a given cutting or rock chip are even smaller. The rock cuttings get all mixed up on the way from 5,335 feet to the surface, so all the cuttings might not be from the oil zone. These rocks were tight; they had no permeability so the carbon tetrachloride would have a difficult time releasing the oil from the rock. All of these factors weighed against obtaining a free-oil cut, that is, an oil release from the rock. Thus, a miniscule sounding of only 2 percent held great meaning.

Measured by past experience, these shows were very good and their persistence had become dramatic. In short, the situation was hard to believe. During my well-sitting experience in West Texas, a premier oil province, I never encountered anything like our well-site geologists were reporting. West Texas shows were subdued, even in very good producers, as compared to the Prudhoe Bay shows. Only twice, and both times in the same oil field in Texas, did I ever see an example of free oil in a dish of well cuttings. And this occurred without the help of carbon tetrachloride.

The reports were beginning to come in akin to someone having a hot streak at the craps or blackjack tables at Las Vegas, frequently referred to as being "on a roll." Here in northern Alaska, ARCO was on a roll. Every couple of hundred feet the intensity of the oil shows increased. At 6,410 feet, the fluorescence went to 100 percent, not unheard of, but certainly unusual, and the quality of the rock improved somewhat to a sandy siltstone. At 6,850 feet, the shows changed to an even more spectacular level. The engineer noted the sample had bright yellow fluorescence, good oil cut (by carbon tetrachloride), and 90 percent dark brown stain.

Siltstone is a very fine-grain rock type and is a reservoir in many oil fields, but not a high quality one. When a siltstone becomes sandy, the quality improves and is more likely to have enhanced porosity and permeability. An already good situation was looking better, but potential reservoir was still not encouraging. Nonetheless, the intensity of the oil shows had created so much curiosity that we decided to run a drill stem test, which meant contacting the drilling engineer.

When the need arises for a special assessment, such as taking a drill stem test, the head geologist speaks to the drilling engineer. Bill Congdon, an ARCO field drilling engineer or supervisor, was ARCO's representative on contract drilling rigs. His specific responsibility was to make sure the company's interests were protected. The day-to-day drilling was the responsibility of the tool pusher, the contractor's senior supervisor on the rig. Sometimes the company engineer was referred to as the company tool pusher. The ARCO drilling engineer is the person to whom the ARCO well-site geologist directs requests to stop drilling, to circulate the bottom hole samples, and look for shows, take a drill stem test or run logging surveys. Bill Congdon or his alternates, Jim Keasler or Joe Mann, were the Prudhoe well's drilling engineers. They in turn would work through the tool pusher to accomplish any given task. Under these circumstances, the company drilling supervisor exercised more direct control over the well than usual, due to occasional increased risk. When the drilling crew performs any operation at the request of the ARCO drilling engineer, the rig time is billed at an agreed upon hourly rate. Bill's dual responsibilities were to watch costs but get the well evaluated.

Artist-geologist Marvin Mangus's painting *Plane tabling (surveying) in the Brooks Range.* Marv is seated and George Gryc is at the instrument.
Photo: John Sweet

1

2

3

4

5

6

7

8

9

1

1. Umiat c. April 1949. Athey wagon tracked trailer to transport fuel drums. These vehicles had low ground-bearing pressure when the tundra thawed. The tracks also helped to smooth the way going cross-country. *Courtesy Marvin D. Mangus Photos*

2. First Atlantic Refining Co. field camp just east of the Canning River in what is now the Arctic National Wildlife Range. It is the general area of the Marsh Creek anticline, which is a prospective oil area within ANWR. The Sadlerochit Mountains are in the near left background. This was also Leffingwell's former stomping ground in 1907–1912. *Courtesy Marvin D. Mangus Photos*

3. USGS field camp during NPR4 exploration on the Chandler River in 1950. *Courtesy Marvin D. Mangus Photos*

4. Marvin Mangus in a military tracked vehicle, *Weasel,* at a rock survey cairn on the divide between the upper reaches of the Colville and Utukok Rivers on the north side of the Brooks Range and the Noatak River on the south side, the first such occurrence. *Courtesy Marvin D. Mangus Photos*

5. Chandler Lake with low sun shadows, and site of USGS field party in 1948. This lake was a favorite camping spot though the years for geological field parties and Eskimos alike. *Courtesy Marvin D. Mangus Photos*

6. Bellanca bush plane on Chandler Lake 1948. This 1932 vintage aircraft is believed to be still operating in Alaska. *Courtesy Marvin D. Mangus Photos*

7. Geophysical camp at Driftwood anticline 1950. Note the freighting sled, left center. Buildings are mounted on sleds. *Courtesy Marvin D. Mangus Photos*

8. NPR4 drilling rig at Gubik anticline in1950. *Courtesy Marvin D. Mangus Photos*

9. Fueling Cessna 195 in Umiat in 1949. Note the bucket fitted with a chamois for filtering unwelcome water from the fuel. *Courtesy Marvin D. Mangus Photos*

10. Food cache barrels at Umiat waiting for the two Norseman planes to fly them to their destinations, an annual activity for Marv Magus through most of the NPR4 field seasons. *Courtesy Marvin D. Mangus Photos*

11

12

13

14

15

16

17

18

20

21

22

11. Gar Pessel, Richfield geologist, near Elusive Lake and the Ribdon River, July 1962. The Bell G-2 Helicopter was in general use by field geologists at that time. *Photo: Charles "Gil" Mull, Consultant, Santa Fe, New Mexico*

12. Richfield geological camp on Chandler Lake, after an early summer snowstorm, June 24, 1964. *Photo: Charles "Gil" Mull, Consultant, Santa Fe, New Mexico*

13. In the haste for the need for a confirmation to the Prudhoe discovery, ARCO–Humble enlisted the services of a C-130. It took much negotiation to get one released from the military, so this was the first commercial use. Here, the next-to-the-largest Caterpillar tractor is unloaded. Marv Mangus is watching in the foreground. *Photo: Charles "Gil" Mull, Consultant, Santa Fe, New Mexico*

14. "Cat" train moving rig from ARCO–Humble Oil *Susie Federal* No.1 to the ARCO–Humble Oil *Prudhoe Bay State* No.1, in the spring of 1968. *Photo: Charles "Gil" Mull, Consultant, Santa Fe, New Mexico*

15. ARCO–Humble Oil *Prudhoe Bay State* No.1, all alone, February 1968. *Photo: Charles "Gil" Mull, Consultant, Santa Fe, New Mexico*

16. Flare from drill stem test no.2, ARCO–Humble Oil *Prudhoe Bay State* No.1, on December 27, 1967. This test was from the Sadlerochit formation. The test was all gas, no oil, and the test tool stuck in the well and was never recovered. This required that a new hole be drilled to isolate the equipment, which could not be recovered. *Photo: Charles "Gil" Mull, Consultant, Santa Fe, New Mexico*

17. ARCO--Humble Oil *Sag River State* No.1, in the spring of 1968. *Photo: Charles "Gil" Mull, Consultant, Santa Fe, New Mexico*

18. A large example of a large oil-stained rock from the Arctic National Wildlife Area. *Photo: Charles "Gil" Mull, Consultant, Santa Fe, New Mexico*

19. This the Atigun Pass area, the route Alyeska Pipe Line Corp. selected for the passage through the Brooks Mountain Range. The view is toward the south. *Photo: Charles "Gil" Mull, Consultant, Santa Fe, New Mexico*

20. A section of the Alaska Pipe Line. It is raised to eliminate heat transfer to the tundra, which might cause melting and destabilization of the line. It is built in a zigzag manner to accommodate potential damage from earthquakes. *Photo: Charles "Gil" Mull, Consultant, Santa Fe, New Mexico*

21. Midnight sun North Slope of Alaska. *Photo: Charles "Gil" Mull, Consultant, Santa Fe, New Mexico*

22. Prudhoe Bay Sunrise. *Photo: Charles "Gil" Mull, Consultant, Santa Fe, New Mexico*

23

24

25

26

27

28

23. Production module, September 1973. These modules were built to connect and are the place where the oil, gas, and water are separated and sent on their way — oil to the Alaska Pipeline, gas for fuel with the rest injected into the producing formation to help maintain reservoir pressure, and water injected around the edge of the oil field also to help preserve reservoir pressure. To save labor costs, these modules were fabricated in Tacoma, Washington and then barged to Prudhoe Bay. Sea ice usually clears away from the coast for two weeks about September 1. *Photo: John Sweet*

24. Special tractors ("creepy crawlers") were designed to drive under the modules, raise them with hydraulic jacks, trundle them off the barge, and then drive them to their respective locations for final fabrication. The very large crawler tracks are visible under this unit. *Photo: John Sweet*

25. Joints of pipe in the pipe coating yard at Prudhoe Bay. Joints are 48 inches in diameter and 40 feet long. *Photo: John Sweet*

26. H.C. "Harry" Jamison briefed the governor on the science we used to justify drilling. Governor Hickel was accompanied by Commissioner for Department of Natural Resources Phil Holdsworth, and Division of State Lands Chief Roscoe E. Bell. *Photo: John Sweet*

27. A more comprehensive aerial view of a geophysical camp. *Photo: Charles "Gil" Mull, Consultant, Santa Fe, New Mexico*

28. Fishing trip with wheels, c.1968. Left to right, all but the pilot are ARCO employees: Louis F. Davis, VP North American Producing, John M. Sweet, Exploration Manager, Alaska, Pilot E.M. "Mo" Benson, VP Oil and Gas, Thornton F. Bradshaw, President, H.C. "Harry" Jamison, Alaska Manager. *Photo: John Sweet*

29. 400 miles of the Alaska Pipeline pipe. It was shipped here to save the overland shipping costs. *Photo: John Sweet*

30. Unloading the pipe for the Alaska pipeline at the Prudhoe Bay dock in 1970. *Photo: John Sweet*

31. Alaska pipeline at the Delta Junction river crossing. *Photo: Howard and Jean Wright 2007*

29

30

31

32

33

34

35

36

37

38

39

AtlanticRichfield—Hum
SAG. RIVER STATE NO.
NW/4 Sec. 4-T10N-R15
Drilling Permit NO. 6

32. 14,500 joints of pipe in the pipe coating yard at Prudhoe Bay. Half of the 800 miles of the pipe line was shipped directly to Prudhoe Bay because that half of the pipe was used to carry oil from Prudhoe Bay to central Alaska where it joined the pipe that had stretched north from Valdez, Alaska. If my arithmetic is correct, that would be 2,212 football fields. *Photo: John Sweet*

33. The tracks of a truck whose driver should have known that he should not do this. *Photo: John Sweet*

34. An early road that is undergoing re-vegetation. *Photo: John Sweet*

35. Tundra soil and moss, 1972. *Photo: John Sweet*

36. "Rollagon," with giant tires to avoid damaging the tundra, during summer of 1973 to retrieve debris from past geophysical activity. *Photo: John Sweet*

37. Shublik and Sadlerochit Mountains to the south of Atlantic's 1963 summer camp, and the site of Ernest Leffingwell's fieldwork during 1907–12. *Photo: Marvin D. Mangus Photos*

38. Winter drilling, December 1974. This is the time of total darkness. *Photo: Charles "Gil" Mull, Consultant, Santa Fe, New Mexico*

39. Robert O. Anderson and sons on a visit when the *Sag River State* No.1 was testing to determine the extent of the oil field. July 1968. *Photo: John Sweet*

40

42

44

45

46

40. The graben went right under this house. A tough way to get a basement! *Photo: John Sweet*

41. This is a key photo in that it shows a great percentage of the Turnagain devastation. McCollie Drive, generally east to west, extends from the left upper edge of the scene with the Norton's (whose experience is included in the text) house fourth from the upper left edge. Turnagain Drive intersects McCollie to the right of center and one can see its ruminants in the debris extending toward the Cook Inlet. Note the red roof of submerged home at about what would be the former end of Turnagain Drive on the bluff of the Inlet. There was only one street parallel to McCollie between it and the Inlet. *Photo: John Sweet*

42. Part of the Government Hill School in the graben. One can imagine the terror of the situation had school been in session. *Photo: W.R. Hansen USGS Alaskan Earthquake no. 67*

43. Many homes were shredded, but others maintained various degrees of integrity, some to the extent that they were salvaged. *Photo: John Sweet*

44. A very catastrophic part of the quake hit "downtown" Anchorage along 4th Avenue generally, but specifically in this block just east of the Anchorage Westward Hotel, which appears in the background. The fault, which caused this, is precisely in this area between these down-dropped buildings and the edge of the street. The fault turned in at the far end of this street and went between them and the hotel. It was theorized by some that the deep footings of the hotel stabilized that area, isolating it from the faulting. *Photo: John Sweet*

45. This substantial I-beam buckled and broke in the Southwest corner of the McKinley High Rise, at eye level. The building still stands. *Photo: John Sweet*

46. The L Street Apartments really felt the strain. It became known as "Epoxy Tower" because much epoxy was used to mend the cracks. *Photo: John Sweet*

47. This is a Parhelion. Leffingwell's sketch (figure 6, page 54) of this in USGS Professional Paper 109 designated by letters the different colors he observed when he saw it. *Photo: Clay S. Turner, Roswell, Georgia*

48. Oil field north of Lander, Wyoming. This is a color version of figure 17.1. *Photo: John Sweet and Jack Nason*

49. An excellent view of the type locality of the Sadlerochit Group of rocks with the formations and members identified. Except for the Ledge Sandstone Member, they are shale equivalents in approximate age to the massive sandstone and conglomerate in the subsurface, which contain all the oil at the Prudhoe Bay oil field. *Photo: Charles "Gil" Mull, Consultant, Santa Fe, New Mexico*

50. I have mentioned anticline, the upward arching of rocks, many times. Shown here is a monocline. The rocks in the upper left of the picture are nearly horizontal. If one could trace the same rock layer or zone from there to the very large white flatirons in the center of the white rocks near the center of the picture, they appear to be dipping sixty degrees from horizontal and are plunging into the earth. This picture was taken from route U.S. 40, near the western border of Colorado. If these same rocks plunged into the earth off the left side, out of the picture, but were visible, the view would be looking at the right of an anticline. *Photo: John Sweet*

48

49

50

At about the time Marvin Mangus began looking for the ARCO site engineer, he appeared. Marv, the well-site geologist for this section, explained to the drilling engineer what he wanted done: the well had to be tested. Bill Congdon would tend to the details of preparing for the drill stem test. The drilling crew went into action to condition the hole and pull the drill, then run it back in the hole with the drill stem test tool. Conditioning the hole translates as circulating the drilling mud for several hours. This builds up a protective skin on the wall of the borehole, making it a safer place in which to work. Since the process would take sixteen to eighteen hours, I seized the opportunity to go from Anchorage to the well site to watch the test, on December 8, 1967. I ended up observing the whole procedure from the opening of the test tool to pulling the drill pipe and seeing the recovery of oil.

A DST is a temporary testing method using the rig drill pipe. The objective is to isolate a zone or rock with an oil show to determine if oil will flow from the rock formation into the drill string. A tool, a rubber packer about four feet long, is attached at a location above the oil show. This rubber packer is placed in the bottom of the drill pipe so it will be above the zone with the oil show. With the bottom of the drill pipe at the bottom of the hole, the weight of the drill pipe above the packer will cause the packer to expand against the conditioned wall of the hole. This forms a plug so the hole below it is isolated. There is another tool in the drill pipe with a valve, which can be opened by turning the drill pipe so many revolutions to open and close it. Usually the crew will fill 1,000 or 1,500 feet of the drill pipe with water. The purpose of this is to subdue an unwanted high-pressure entry into the pipe from the rock formation. It is a measure of insurance to help prevent an uncontrolled flow of oil or gas. Opening the valve in the tool allows whatever fluid — gas, oil or water — that is in the rock formation to flow into the drill pipe. An ideal test occurs when oil comes quickly to the surface.

A DST was taken during the interval from 6,876 to the bottom of the hole at 6,998 feet. When the tool was opened, there was a moderate flow of gas at the surface and a recovery of 1,500 feet of oil in the drill pipe. In the South 48, this would have been something to get excited about. On the Arctic Slope, it was certainly encouraging. It indicated that the good shows we had been logging from the drill cuttings had been authentic, not an aberration. In fact, the shows of oil had been so good it was difficult to believe. Usually, anything that seems too good to be true turns out to be just that — but not here. Mangus and Penttila had seen excellent shows of oil. The rocks were just too tight and not good reservoirs.

When examining the Schlumberger log (Figure 20) over the section of the well that was tested, the wiggly line on the left side of the log is characteristic of shale or siltstone. The slight spike at the 7,000-foot depth is indicative of sand, or more likely sandy siltstone. The best part of this thin zone is probably six to eight feet thick. This is a good opportunity to remind readers that looking at maps and logs is a world apart from walking along a rock outcrop or standing beside an outcrop. That little blip looks rather insignificant in the long shale section on the log, but if it was eight feet thick and one was standing by it, it would be impressive. Think what 200 feet of Prudhoe Bay oil sand would look like if you were standing beside it.

The oil and gas recovered from the drill stem test probably came from this thin sand zone, seen in Figure 20. The Prudhoe Bay oil zone, or the "pay" — yet to be found deeper than the test seen in Figure 20 — was dramatically thicker; and because of the good quality of the sand reservoir in the deeper zone, it was also found to have excellent porosity and permeability. This was discovered in the confirmation ARCO–Humble *Sag River State* No.1 (Figure 22, page 240).

Immediately below the tested interval at 6,998 feet we decided to cut a diamond core, which yields layers of various thickness of rock, four inches in diameter, enabling a better look at the rocks than the microscopic chips from ordinary drilling. We recovered mostly

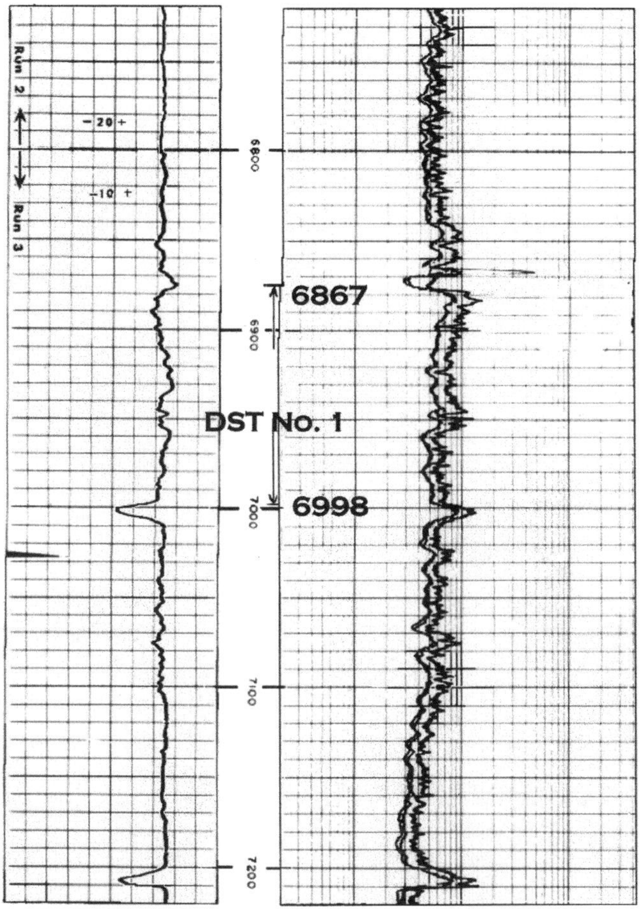

Figure 20: The Schlumberger log of first oil recovery. There had been so many gas and, increasingly, oil shows that even in the absence of good reservoir rocks we decided we had to take a drill stem test (DST) on Dec. 8, 1967. This first test was from a Cretaceous interval from 6,876 feel to 6,998 feet. Gas came to the surface at the rate of 500,000 cu. ft. per day, and then declined to 50,000 cu. ft. per day. The crew found 1,500 feet of oil in the drill pipe when they pulled the tester out of the hole. The test is plotted on a Schlumberger ("Schlumberjay") electric log. The curve on the left of this log typically is indicative of shale when it is close to the center column, as it is here. The two small "spikes" to the left at 7,000 and 7,208 feet may have been thin sand streaks. The test 6,876 to 6,998 may have just scratched, or been close enough to the streak at 7,000 feet for it to have yielded the oil. This log is another example of how things are some times minimized by the tools. The spike at 7,000 feet may be eight feet thick. That doesn't look like much and would not make much of an oil field. On the other hand, if one was standing by an outcrop of rock ten feet thick, it would be much more impressive. After Rutledge, Gene, Prudhoe Bay Discovery! *Wolfe Business Services (Copyright) 1987 reprinted by permission of Gene Rutledge.*

shale and siltstone with thin and very fine sandstone intervals. The core was darkly oil stained with visible free oil and good oil odor.

From 3,300 feet to 7,050 feet there had been more than 3,750 feet of oil shows; likewise, a significant amount of oil was recovered in a drill stem test. Although not commercial here, this was a harbinger for the future. In future wells, oil was found throughout the geologic column (Figure 28) from near the surface to the base of the Mississippian Lisburne limestone at 11,000 feet.

When asked if the days drilling the discovery well were exciting, my comment usually is, "Not particularly." It was like watching a tree

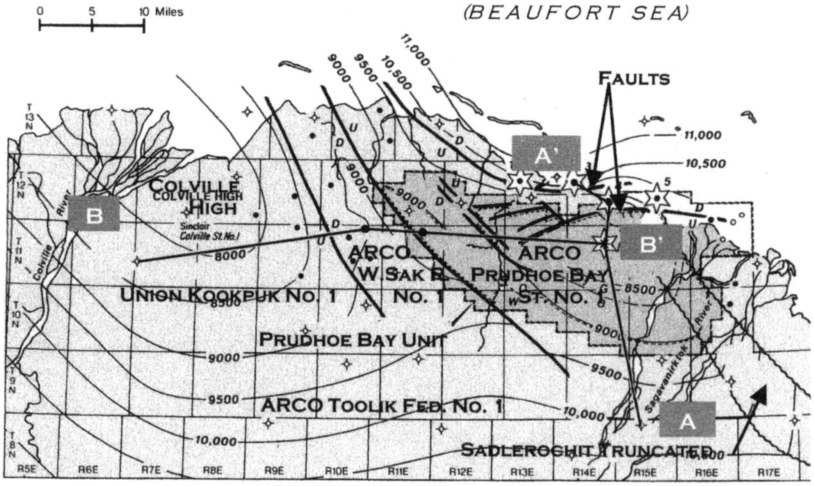

Figure 21: Sadlerochit Formation, the main reservoir. Contour Interval 500 feet. The cross-sections Figures15 and 16 show some faulting, and the zone of truncation of the Sadlerochit have been labeled for location on Figures 14 and 21. The map covers roughly the same area as Figure 14. The geology of the two maps is essentially the same, but several things will make it look quite different to the lay reader. The scale of Figure 21 is smaller than the scale of Figure 14 to the extent that Figure 21 covers sixteen times more land per map unit. The contour spacing is 100 feet on Figure 14 and 500 feet on Figure 21 so it has one-fifth of the contours. Figure 14 (page 169) was based on seismic mapping. Figure 21 is based on much more seismic and the precise information from many wells drilled during the ten years since the discovery.

After Jamison, H.C. et al. Giant Oil and Gas Fields of the Decade 1968–1978. *AAPG ©1980 reprinted by permission of the AAPG whose permission is required for further use.*

grow, because an interesting picture unfolding while sitting a wildcat well never happens fast enough. With oil wells there are occasionally problems with the drill string getting stuck, losing mud circulation because of the bit encountering fractured or very porous rock, losing the drill pipe in the hole, or mechanical failures of the equipment. Invariably, it seems like these things happen at inopportune times, such as with good shows, or a drilling break, when the indicators become interesting or when the drilling approaches an important geologic formation. More than a few times during the drilling of the *Prudhoe Bay State* No.1 discovery well, the drilling crew experienced any one of these. Think of it this way — what has been presented so far in this chapter has taken a short time to read, but the time period covered extends from April 22 through December 9, almost eight months.

On this very same subject, in an interview in 1979, H.C. Jamison said,

> The thing was very gradual. A lot of people have said, 'Gee, weren't you excited, wasn't this a big climax?' The answer is really 'no' because it wasn't. It was a gradual thing and the implications were primarily gas. For warranting the well, the extrapolation into the magnitude of a ten-billion-barrel oil field was sometime in coming. I guess it was dramatic and I guess I was enthusiastic, but I wasn't turning handsprings.

The excellent quality of the shows of oil had actually spooked us; they were encouraging beyond our experience. This led to caution rather than excitement. Many were cautious because most petroleum geologists know what it's like to have had their high hopes dashed. In retrospect, there was much more excitement after the confirmation well. *The Sag River State* No.1 proved conclusively that we had found the proverbial elephant-size field. It was drilled rapidly with few problems, which is unusual, and it was begun so quickly it was virtually concurrent with the *Prudhoe Bay State* No.1.

The Sadlerochit Formation

Following the drill stem test 6,998, in the *Prudhoe Bay State* No.1, the next 1,200 feet of hole was somewhat the same with oil shows, but no reservoir rock. Just below 8,200 feet, drilling penetrated the Sadlerochit formation. Again, indicators from the cuttings were very encouraging and cores revealed sandstone, fine to medium grained, but also occasionally bearing conglomerate streaks with good yellow fluorescence, fair oil cut and good oil stain. These rocks were potentially better reservoirs, and a good sign, because it was the lack of reservoirs that had been plaguing us.

In response to how powerful the chromatograph shows had become, ARCO geologist William C. Penttila, recalls,

> Our procedure in examining the drilling mud and cuttings was to place a certain measure of drill cuttings and mud in a blender, agitate it for a very short duration, then read [with the chromatograph] the resulting content of the blended mixture. We very soon abandoned this procedure while drilling the Sadlerochit gas and oil reservoir because the resulting gas content was so high the blender would blow its top when the switch was thrown to take a reading. Mud and cuttings would be splattered on us and everything in the mud logging trailer.

The Sadlerochit Group is Permian–Triassic geologic age (Figures 21 and 25). For brevity and convenience, only the name Sadlerochit will be used to identify the prolific oil producing zones in the *Prudhoe Bay State* No.1 discovery. The samples, oil shows, and gas chromatograph all continued to yield favorable oil information (Figure 19). On December 27, 1967, we decided to take the second drill stem test from 8,410 to 8,593 feet, and it flowed gas (not what we needed or wanted!) at the rate of 1.25 million cubic feet per day. This was the proverbial good news, bad news situation. The good news was that we had a successful hydrocarbon test; the bad news was, of course, the gas. And while gas is not what we wanted, it was far better than water. Humble Oil and Refining geologist Gil Mull recalls, "As soon as the test tool was opened, there was an immediate strong flow of gas to the surface, which was diverted to the side of the rig. This thing blew with a roar that was something like the rumble of a jet plane overhead that you could not only hear but feel through your feet." And then, to add to our disappointment, the drill pipe stuck when the driller started pulling out of the hole with the test tools. This was one of those unwanted and frustrating delays, just when interest in the well had again increased. The drillers had to divert from drilling and looking for oil, to fishing the lost drill pipe and test assembly out of the hole. Although we had not found oil, the test was encouraging and we were eager to see what lay below. Trees grow slowly.

Soon we realized that Murphy's Law had taken over — if something can go wrong it will, especially on wildcat wells. The information was getting exciting and the whole crew was at the threshold of learning what the data meant, and now we had to go fishing! (Not the fun kind.) We had the DST going favorably, and now the drilling crew had to fish the test tool out of the hole before the rig could be used for anything else. Fishing for oil field tools lost in the hole is frustrating, slow and costly, yet sometimes simple and successful. Other times, Murphy wins, and the entire hole, or parts of it, may have to be abandoned. None of the fish ("tools," "junk," "iron") were recovered, so the *Prudhoe Bay State* No.1 had to be plugged at 7,800 and redrilled to 8,593 feet. This took thirty-six days, until February 1, 1968. Although this wait was just over a month, with anticipation rampant it seemed like an eternity. Emotional frustration blanketed the project during this delay.

When drilling resumed, we cut several cores in the 8,100–8,500-foot interval (Figure 22). The cores had very good shows; the porosity was, incredibly, in the 20–27 percent range, which is like a sponge for an oil reservoir. Porosity gives the rock the voids it needs to hold oil. Ordinarily, 10–15 percent makes a

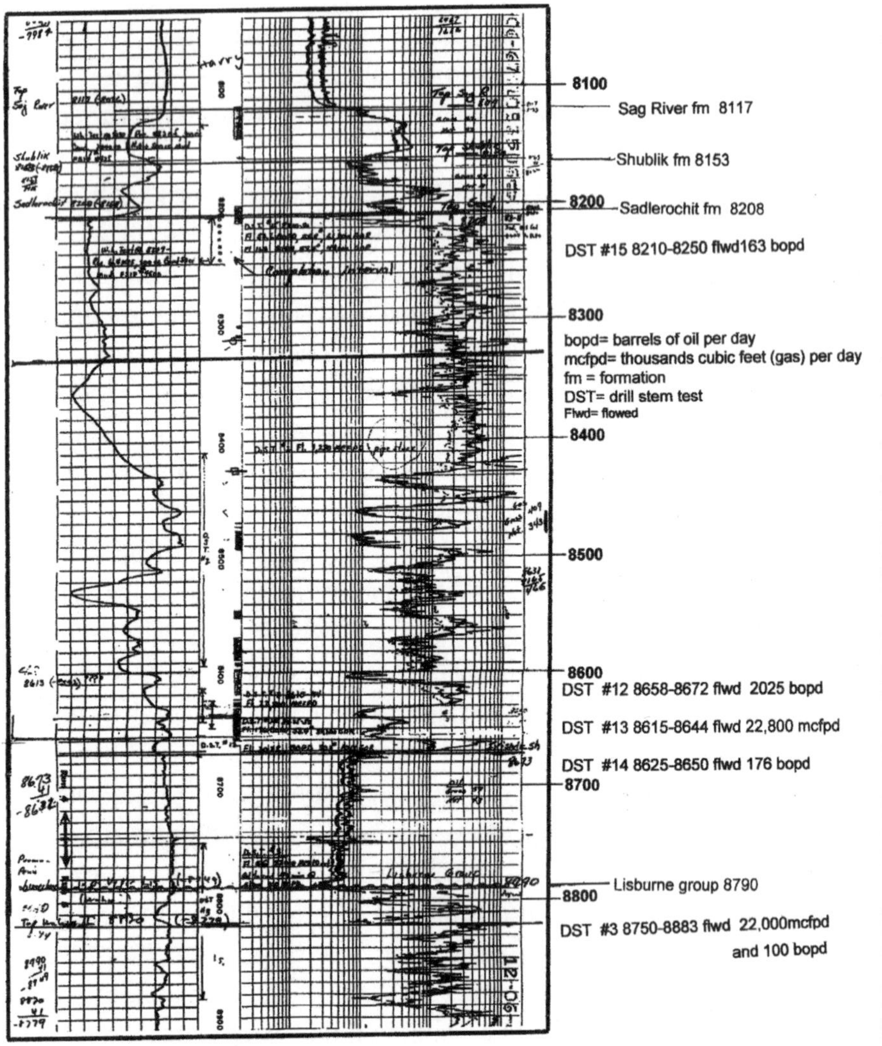

Figure 22: The Second Oil Recovery and Bonanza. This is a portion of the ARCO–Humb Prudhoe Bay State *No.1 Schlumberger well log with the geologist's original notations. Comparing this to Figure 20, there is a small segment of the shale line at the top of the log in the left hand column. Then at the depth of 8,210 where the Sadlerochit formations begins, the log curve excursion to the left indicates sandstone and conglomerate for most of the next 500 feet. The opposite curves on the right indicate oil. The drill stem tests have been added for the reader's benefit. At this scale the geol ogist's comments are difficult to make out for the lay person, but they are present throughout the log.*

bopd	*= barrels of oil per day*	*Fm*	*= formation*
mcfpd	*= thousand cubic feet (gas) per day*	*DST*	*= drill stem test*
Flwd	*= flowed*		

geologist happy. The rock types were sandstone and conglomerate. In addition to the descriptions by the well-site geologist, they were described for the characteristics of depositional environment by Kirby Bay in the Geoscience Group in Dallas. Dr. David Hite, a geological expert in clastic (sand, shale, conglomerate) deposition prepared an even more comprehensive study.

The Sadlerochit oil reservoir is unique, complex and difficult to describe. I will make it as simple as I can, having been coached by Dr. Hite. The producing rock unit, the Ivishak Sandstone member of the Sadlerochit, was deposited in a sea with strong currents and a steep bottom, called a fan delta. These deltas tend to evidence great continuity. Other deltas, like the Mississippi Delta, change abruptly over short distances and have fine-grained rocks, which are not good oil reservoirs. A fan delta can be visualized as a conical shaped body of sand and gravel with relatively little fine-grained rocks, like shale. The Prudhoe fan delta built seaward from a highland source area north of the present oil field, which is now the Beaufort Sea.

Anyone who has traveled in the American West has seen, perhaps unknowingly, fans or cone-like features at the mouths of canyons on mountain range fronts. These fans spread out into the valley and sometimes merge laterally to form a system along the mountain front. Imagine that the valley is a large lake or sea and that the fans are being deposited in the water. That is a fan delta. The great bulk of fan deltas, Prudhoe Bay included, are comprised of sand (sandstone) and gravel (conglomerate); if the climate is wet enough, the sandstone and conglomerate rocks are clean. This occurs because lots of water will carry the fine silt and clay particles farther into the sea and will not plug up the porosity and permeability of the sandstone and conglomerate rocks.

At Prudhoe Bay, other depositional environments associated with fan deltas also are present. They are marine bars, river mouth bars, braided sandstone river channels and meandering conglomerate filled channels. Marine bars are the first of the sequence of deposition into the sea, followed by braided channel deposits on top of the bars, and finally, the fan deltas on top of the braided channels. The bars have the greatest seaward extent and the fans the most limited. The dominant feature at Prudhoe Bay is the fan delta. It is the reason the continuity of rocks between the *Prudhoe Bay State* No.1 and the confirmation *Sag River State* No.1, seven miles southeast, is so good.

The permeability of the rock was measured as high as several darcies. This is a measure of the flow capability between pores within the rock, so that when you find oil, it will flow through the rock to the borehole and can be recovered. Most of us on the site had experienced rock permeability measured in tens of millidarcies (1,000 millidarcies is one darcy). The flow capability was beyond our collective experience. It is the same amount of porosity that was seen in some early U.S. discoveries and in Middle East oil fields.

Usually, cores come from the core barrel in the shape of three- or four-inch diameter cylinders or thinner biscuits of rock. The sandstone and conglomerate being cored might look not so different to the untrained eye than a core cut in a sidewalk. The induration, or hardness, of a core is usually indicative of what one might expect from a laboratory analysis. Hard usually means tight and potentially unproductive; friable usually means some porosity and permeability will be present. In P*rudhoe Bay, Discovery to Recovery* by Gene Rutledge, Gil Mull, then a Humble geologist, describes a core of the Sadlerochit, the producing formation.

> What came out instead of solid rock was a pile of disaggregated loose sand, gravel, and oil — some of which immediately ran through the rig floor into the rig cellar. With this, in spite of the attempts at secrecy, most of the drilling crew had a pretty good idea of what was being found.

Not only had many been privy to the secret, but Marvin Mangus

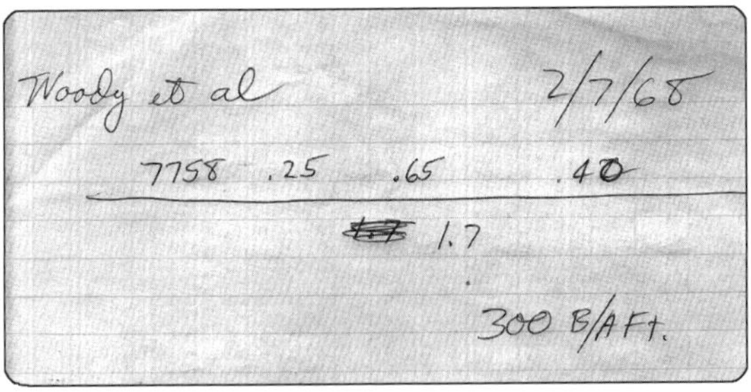

Figure 23: The First Oil Reserve Calculation, first page. The formula at the top of the page is the basis for calculating reserves. 7,759 represents the volume of oil in an acre one foot deep. The oil is in voids (porosity) in the rock. Here it was 25 percent. The voids also contain water and at Prudhoe it was estimated at this time to be 35 percent, so 1–35 percent, or 65 percent, of the void is available for oil. All the oil is never produced. Here we estimated 40 percent would be recovered. The product of these parameters is then divided by the formation volume factor, which is the change in volume between the reservoir and at atmospheric pressure. The result is the barrels of oil per acre foot of thickness of oil column, or at Prudhoe Bay, 300 barrels per acre foot in the reservoir.

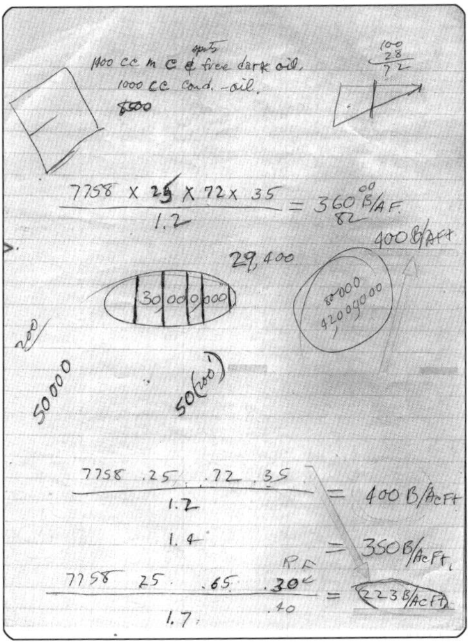

Figure 24: The First Oil Reserve Calculation, second page. The figures at the bottom of the page combine the known parameters of 7,758, which are the barrels of oil per acre, one foot thick, and the porosity of 25 percent. The next parameter is the percent of oil in the reservoir, and the final parameter is the percent of oil that may be produced. The different divisor represents different formation volume factors, which are the changes between the reservoir pressure and ambient air pressure. 400, 350 and 223 represent the different number of barrels per foot in the three different calculations.

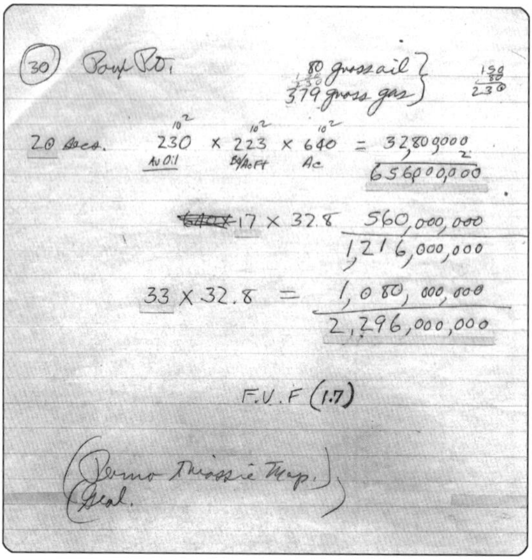

Figure 25: The First Oil Reserve Calculation, third page. On this page John used averages from Figure 24 of oil pay thickness of 230 feet, barrels of oil per acre foot of 223 and 640 acres per section, then the applied number of sections to arrive at three numbers of reserves. Twenty sections proved equal 656,000,000 barrels of reserves, 17 additional sections of 560,000,000 barrels of potential barrels of reserves, and 33 more additional sections of 1,080,000,000 barrels of possible reserves.

The total proved, potential and possible reserves were 2,296,000,000. Although John didn't know at the time, the field ultimately covered 200 sections. The average of oil pay thickness and per acre feet of recovery also had to be increased about 40 percent to combine with 200 sections to make the 10,000,000,000 original reserves.

also learned a few years later that a roughneck (drilling crew person) had stolen a piece of the core, a particularly good specimen of the Sadlerochit formation. Marv was working as a consultant for Forest Oil Company on a wildcat well to the south of the Prudhoe Bay area when the culprit confessed. He promised to send the sample to Marv, but it never arrived.

In the redrilled hole, we cored most of the interval from 8,550 to 8,711 feet and the bottom of the Sadlerochit was at 8,673. On

February 7, 1968, when we received the first measurements from the core analysis, I made what was the first authoritative oil reserve estimate of the Prudhoe Bay potential (Figures 22, 23, 24 and 25). These measurements, along with what we had learned from the position of the well on the structure and the geophysics, made it possible, within limits, to calculate the likely reserves. We learned from the core analysis that we had about sixty feet rock formation with oil (oil pay) at the very bottom of the formation.

I consulted with Woody Kingsbury, our district petroleum engineer, to make sure I hadn't overlooked something or miscalculated. At this point, everything became super secret, because the information we now had was of greater proprietary value than any of us could have ever conceived.

These original reserve calculations (Figures 23, 24, 25) indicated that the *Prudhoe Bay State* No.1 had discovered a minimum of 656 million barrels of oil. This calculation meant if there was no more oil found than the sixty feet of rock formation with oil (oil pay or oil column) in the discovery, the area of the production would be under twenty sections of land (640 acres each) and the reserves would be 656 million barrels.

I estimated a "what if" of another fifty (17+33) sections, for a total of seventy (20+17+33) sections. To do this, I had to assume the oil column would be thicker than the sixty feet already discovered. There was every reason to believe that there would be a thicker oil column because no indication of water beneath the oil surfaced. The unknown factor was how much more oil pay. I used an average estimate thickness of 230 feet. The results of my "what if" exercise increased the potential discovery to 2.3 billion barrels of oil. As it turned out, that seventy-section estimate was very conservative, because the Prudhoe Bay field ultimately covered about 200 sections with 13 billion barrels of oil. I remember thinking, after having made the calculation, "no one is going to believe this." In fact, district geologist Don Jessup thought I was off my rocker calling in the estimate to the Dallas staff.

One month later, Woody Kingsbury confirmed that we had seventy feet of gross oil column — ten feet more than the cores had indicated. He further calculated that if there was an oil column 200 feet under 44,000 acres (almost seventy sections), reserves would be between 1.83 billion and 3.66 billion barrels of oil. He arrived at these numbers after reflecting on the information from the coring, Schlumberger electric logs, and an evaluation of what was known at that point. Kingsbury probably used different water content and residual oil saturation values, causing his variation in reserve size. He was uncertain of the validity of the low water and high residual oil factors from the core analysis because these two excellent and important parameters were so much better than his range of experience. Engineers were responsible for reserve determination in ARCO, so I was pleased that their calculations were so similar to mine.

It is a truism that poor oil fields are never as good as one hopes, because of one's desire. One wants oil fields to be good, and thus generally overestimates their potential. Many situations in life are like that. On the other hand, after one has been chastened a few times, and then a really good field comes along, one tends to be too conservative.

In *Prudhoe—A 10-Year Perspective,* Jamison, Brockett and McIntosh summarized the Sadlerochit reservoir. It appears here in table form:

	Thickness	Porosity	Permeability
Upper	200 feet	25-30%	500-4000 md
Conglomerate	40-140 feet	10-20%	500-1000 md
Lower	300 feet	25-30%	250-3000 md

After the well was drilled to total depth, steel casing was put in the hole. Logically, when drilling the tests are taken on the way down.

Conversely, when production testing is done, it is usually done from the bottom up. Figure 26 is an "isopach" (thickness) map of he Sadlerochit formation. It is over 500 feet thick over most of the field.

The Schlumberger logs, similar to Figures 20 and 22, enabled the reservoir engineers to more precisely pin down important parameters, such as the exact zones of oil, water saturation, porosity in the sections not cored, and the oil water contacts. Two different tests taken of perforations in the well casing in the Sadlerochit pinpointed the gas–oil contact in a shale zone between 8,644 and 8,656 feet.

The oil test below is from 8,658 to 8,670 feet. The earlier DST taken at 6,998 was in open hole. Tests from there on were taken with steel casing (pipe) in the hole. The pipe is perforated to let the oil come in. These tests were taken through various restrictions in the flow line, which are called chokes. In the table, you can see how flow rates vary with choke size.

Duration of Test	Choke or Opening in Pipe	Barrels of Oil Per Day
4 hours	½ inch	1740
1 hour	0.625 inch	2118
1 hour	1 inch	2400
¼ hour	2 inches	2415

Testing ended June 3, 1968, and it was easy to speculate that if twelve feet would yield these kinds of rates, ten times the thickness should yield 24,000 barrels per day. This is an oversimplification, but many wells did produce at 24,000 barrels per day, and then some.

What a success! And yet, evaluating the Lisburne formation was the real goal, and that was still to come.

Chapter 14

Lisburne Limestone: The Main Objective

The *Prudhoe Bay State* No.1 was originally proposed and started with the Lisburne limestone of Mississippian age as its primary objective. Those who had studied the geology believed that the Lisburne had the attributes of a good oil reservoir. The fact that ARCO–Humble had discovered a major oil field in the Sadlerochit formation, with more than 500 feet of excellent appearing reservoir rock and good shows, simply enhanced the assessment of the Lisburne. Logically, if the Lisburne was primary to begin with, the Sadlerochit production proved the well was in oil country. There was every reason to be optimistic. Hopes were up and expectations high. At about 8,800 feet the drill bit cut into the Lisburne limestone (Figures 27 and 28). Figure 27 on page 250 is the lithologic log with the symbols and the curves of the Schlumberger log in the Lisburne formation in the discovery well. Figure 28 on page 255 is the Lisburne structure map. It approximates the structure of the Sadlerochit formation in Figure 21, page 236.

Unless limestone reservoirs are in reefs — which by character are frequently composed of a heterogeneous conglomeration of reef skeletal parts — they are more likely to have porosity and permeability compared to bedded limestone rocks.

Layered limestone usually has much lower porosity and permeability than in clastic (sand) rocks because they are precipitated calcium carbonate, which tends to pack tightly and is, therefore, generally hard and dense. (Author's note: Reefs described here are in the context of an oceanic environment, but fossil reefs have also been found thousands of feet below the surface, high in the mountains, and anywhere in between.) It is not unusual for marine organisms, circulating ground water, or erosion to cause secondary porosity, to give it oil reservoir qualities after deposition.

Not long after the well was drilled into the Lisburne formation, it was cored. The analysis revealed porosity ranging from none up to 20 percent, but with permeability of only one millidarcy, quite a come-down from the darcies of permeability in the Sadlerochit. Lower millidarcies mean the rock will not give up oil easily, a bad situation. The formation was substantially fractured (broken by nature with large cracks), which enhances potential productivity. There were good shows of oil in the cores. A level of 20 percent porosity is higher than most limestone reservoirs. The low permeability of the rock was not surprising; it is common in limestone, but nevertheless disappointing with such good porosity.

The oil shows in cores, well cuttings and the mud log were all the evidence the geologists needed for another drill stem test. Test no.3 from 8,686 to 8,883 feet flowed gas to the surface at the rate of 22-million cubic feet per day, plus a small flow of oil, 100 barrels per day. Again, the gas was better than water, but the crew needed more oil volume before we could call it worthwhile.

The drilling continued, and the samples and mud log exhibited an increase of oil shows. More cores between 9,510 and 9,867 feet continued to encourage the geologists, leading them to take two more drill stem tests. Bill Penttila, ARCO's well-site geologist, and Humble's Gil Mull worked cooperatively. Here is Bill's account of a specific test.

> There was an increase in the gas and oil in the drilling mud as indicated by the mud logging equipment and oil showings in the drill cuttings. However, these were much more subtle than those

Figure 26: Sadlerochit Formation Thickness (Isopach) Map.

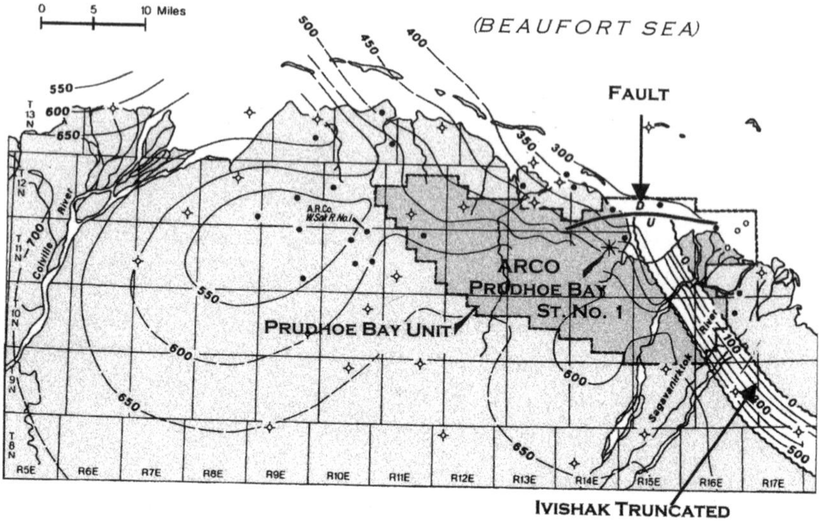

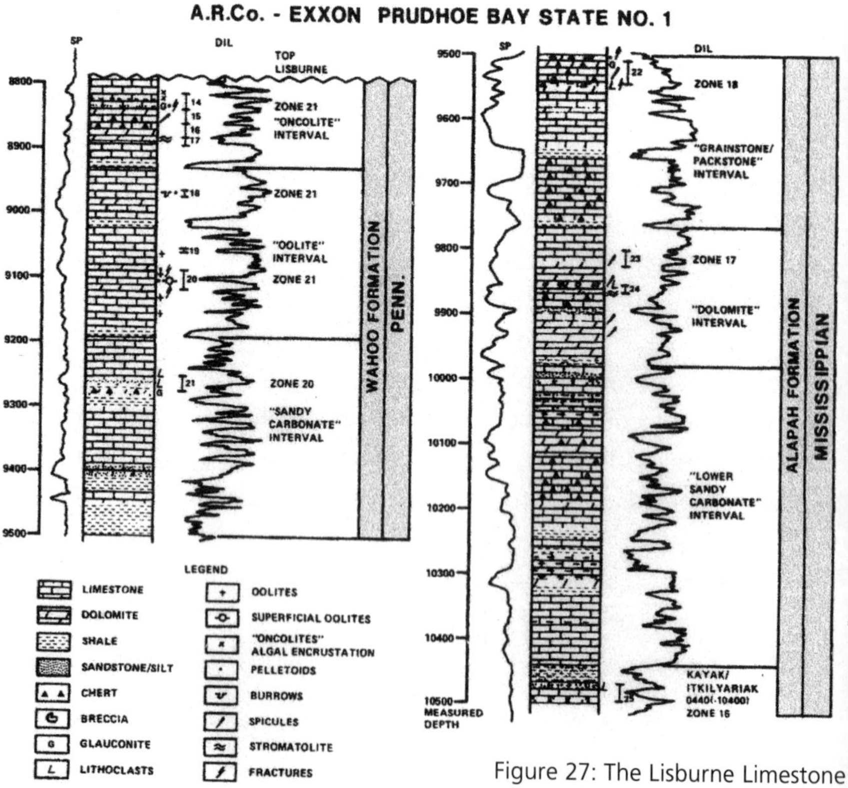

Figure 27: The Lisburne Limestone Zone in the Discovery Well.

shows previously in the Sadlerochit sandstones. After satisfying myself that we in fact had all the signs of a valid showing, and that the interval we penetrated had sufficient porosity to be an oil reservoir, I called for an open-hole drill stem test.

This met with opposition from the ARCO drilling foreman's boss, Benny Laudermilk, who was on location inspecting and evaluating the drilling operation. Benny, for whom I have a lot of respect as a drilling operations person, said he had talked to the derrick man and the derrick man told him that he did not see any oil shows in the drill cuttings.

The derrick man is one of the drilling crew members who usually collect drill cuttings from the drilling mud for the well-site geologist and mud loggers to examine. The derrick man usually has a great deal of experience looking at drill cuttings, and can be quite good at recognizing oil shows. However, derrick men do not have the technical training and tools to

Figure 26: This is an isopach (thickness) map of the Sadlerochit (Ivishak) formation. Contour Interval 50 feet.The truncation zone on the east end of the field, is shown here, and on earlier figures. West of where the truncation begins, the thickness of the Sadlerochit increases to a maximum of 850 feet at the southwest part of the map. The average appears to be 750 feet. Much of the thickest part is outside the producing area. Contours are in feet.

After Jamison, H.C. et al. Giant Oil and Gas Fields of the Decade 1968-1978. *AAPG (Copyright) 1980 reprinted by permission of the AAPG whose permission is required for further use.*

Figure 27: The main objective was the Lisburne limestone. It has produced oil but was insignificant.

After Jamison, H.C. et al. Giant Oil and Gas Fields of the Decade 1968–1978 *AAPG ©1980 reprinted by permission of the AAPG whose permission is required for further use.*

> adequately make the qualitative judgment of oil and gas show, especially if the shows are very subtle. This is one of the reasons for the well-site geologist.
>
> I insisted that we run a test. The procedure for opening the tool to begin the test was for the tool to be rotated a certain number of turns in one direction and a certain number in the opposite direction. I was on the rig floor in thirty- and forty-degree-below-zero weather watching this operation, and observed that these rotations were not in accordance with procedures. The testing continued but at the end of the test it was declared a failure. Benny wanted to resume drilling because there was no oil or gas recovery with the comment 'there isn't anything there anyway.'

ARCO had great drilling engineers, and they rightfully became uneasy if they saw signs of trouble. A failed test is a sign of potential trouble. Remember the Sadlerochit DST back at the end of December, when the drill pipe got stuck at the end of a DST, causing over a month of lost drilling time? This was undoubtedly in the minds of Benny Laudermilk and his drilling engineers, who also liked to find oil, but they sometimes balked at a geologist's request for a drill stem test. The engineers' primary focus is to drill a safe hole in the earth as cost-effectively as possible. Benny Laudermilk was the district drilling supervisor and had years of experience. He was responsible for drilling a good hole. The trouble was that Benny was strongly questioning the wisdom of the geologist. Benny was outside his realm of training by questioning the advisability of retesting the zone. His drilling engineers may have failed in their responsibilities during the first aborted test, if Penttila was correct, and the tool had not been opened according to procedures. The engineers should have been watching that part of the operation. When the temperature is thirty to forty degrees below zero, mistakes come easily, and the possibility exists that Penttila counted the revolutions incorrectly. There is no evidence of personality conflicts, but rather an earnest desire of both men to make the right decision, each thinking their position was the correct one.

Bill Penttila continues.

> I maintained we didn't get a proper test and that we should re-test. We were both adamant in our positions, so Benny insisted we both fly to Pt. Barrow to call our bosses in Anchorage to resolve our standoff. We flew in the King Airplane to Pt. Barrow, but the visibility at the airstrip was so bad that we made six or more attempts to locate the strip, but were only able to catch a couple of fleeting glimpses of the ground.
>
> Eventually the pilot said it was too risky to try any more, so we returned to the drill site. I maintained that we had to test these Lisburne shows, and finally another tester was run in the hole. I watched again to make sure the operation followed the proper procedures.
>
> Upon the opening of the test tool, there were immediate indications of fluid flow into the testing tool by the blow of air followed by gas into the flow line at the rig floor. I went from the rig floor to the end of the flow line [the flow line is directed away from the rig for safety reasons, because in the event of oil or gas flowing, it can be safely set on fire and monitored], which is some distance from the rig, and observed intermediate slugs of drilling mud being blown out the end of the flow line with loud 'whoosh' sounds, shortly followed by slugs of mud and oil. Then green-brown oil [came] with even louder 'whoo-whoo-whoosh' and the slugs coming closer together until there was a continuous stream of oil flowing to the pits.
>
> When I returned to the rig no one would believe me, until some

roughnecks came into the mess hall with the same story. The flow was burned because there was no storage for so great a volume of oil. This was seen by a commercial flight and reported to Anchorage as the rig and camp were burning.

On March 12, 1968, the success of the test validated Penttila's conviction that another DST was justified. The results of the test were dramatic. So much for the derrick man and his observation that there were no shows in the samples. He, too, was outside his sphere of training when he gave this opinion. This was drill stem test no.5, of the interval 9,505–9,825 feet. Gas flowed to the surface in twenty-three minutes, and then oil flowed for two hours and ten minutes at the rate of 1,152 barrels of oil per day, accompanied by 1.3 million cubic feet of gas per day.

Flying to Barrow began to look like a flawed cost-cutting measure, maybe poor judgment, because the flight time was expensive and the rig costs continued. Generally speaking, the drillers would not keep the drill pipe out of the hole too long. While the retesting scenario was being debated, the drilling crew probably re-entered the hole, circulating the mud to keep the hole in good condition. In all probability this delay was more costly than a drill stem test would have been.

If Bill Penttila had not persisted, the course of events probably would have been little affected, but the test was spectacular and important in planning a future course of action. All's well that ends well, as the saying goes.

Things were not going well at all in Roswell, New Mexico where the Dallas staff, led by Louis F. Davis, vice president, had gone for a conference. Customarily, information is passed up through the chain of command. No sooner had the Dallas group arrived (11:30 a.m. MDT) than chairman of ARCO, R.O. Anderson, whose home happens to be near Roswell, tracked Louis Davis down to tell him the well was on fire. A passing airliner had spotted the bright red flame

Figure 28: The Lisburne Structural Map.

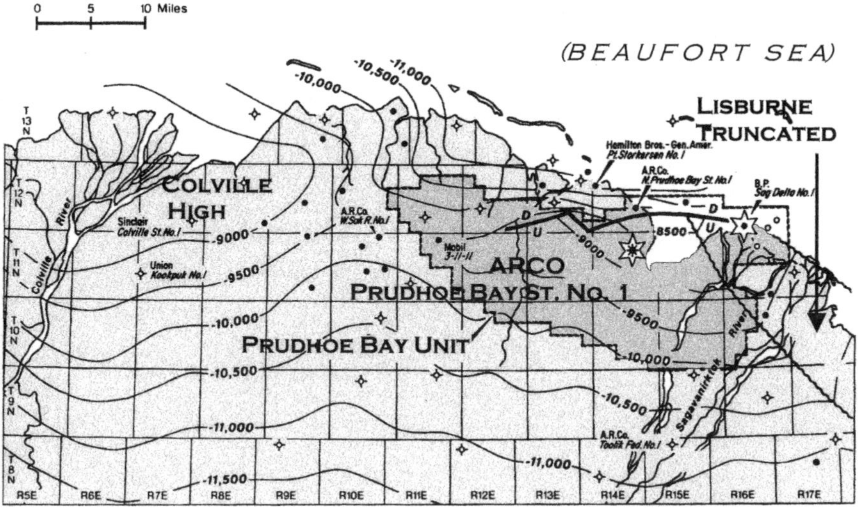

In most of the area it conforms generally to the Sadlerochit, which lies above it. Contours are in feet. The contour interval is 500 feet.

After Jamison, H.C. et al. Giant Oil and Gas Fields of the Decade 1968–1978. *AAPG (Copyright) 1980 reprinted with permission of the AAPG whose permission is required for further use.*

at remote Prudhoe Bay. One way or another, that information had reached the top man, some 4,000 miles away in Roswell.

How did this happen? The answer has been lost with time, but it brings to mind an Alaskan phrase, "mukluk telegraph." Mukluk in the Eskimo language means seal; it also became the name for the boots that the aboriginal people and others wear in northern Alaska. Communications in Alaska are sometimes uncanny, to say the least. People in one area find out rapidly what is going on somewhere else, sometimes hundreds of miles away, yet there is no apparent or obvious means of communication. It appears beyond any obvious explanation but has been named from being carried on foot in mukluks as mukluk telegraph.

Robert O. Anderson unknowingly could have been hooked into the mukluk telegraph. More likely, the pilot radioed to somebody in Anchorage, and the information flooded through other radios, then into the phone system, and then to the newspaper people. The line of communication does not stop there, for the media takes great delight in going to top executives with gossip and scoops.

Meanwhile, in Roswell, Mr. Davis needed answers. In situations like this, Davis would frequently turn to Julius Babisak, exploration manager, to find out what was happening. Julius was expert at providing answers to off-the-wall questions and requests from Mr. Davis.

Switch now to Anchorage. This was what was happening then as Julius learned from Lee Wilson.

> Those of us back in Anchorage waited anxiously for some word from the well, but as usual, there was no response from our radio. We wanted this data right away because this was such an important wildcat, and the "tight hole" status made information very time sensitive. Tight hole is oil-field jargon for very confidential well information.
>
> About noon we received a frantic call from Interior Airways in Fairbanks stating that the rig was on fire. The message had come from the Airways radio at the airstrip. We were shocked. No. We were thunderstruck. As drilling superintendent, I was especially staggered, since it was my responsibility to be sure to run a safe test. I called the district manager, Harry Jamison, at once. He was equally stunned by the news. [During an edit Harry penned, "You bet I was."] A charter plane was hastily arranged for us to go to the fire. It was about thirty below on the Slope. I began thinking about Red Adair and all the problems of a blowout at such a remote site, being in the Arctic at that.
>
> This second call, which seemed to us to be after several hours

> of hectic action, was really only a half hour later. The rig was not on fire; it only appeared to be. The radio operator in the shack at the strip had a view of the rig. An Alaskan, he was unfamiliar with the well site goings-on. He didn't know what a DST was, let alone, observe one. At the rig the short line that was strung to carry the gas or oil, if any, away from it was in a direction exactly, inadvertently, toward the radio shack. The DST was spectacularly successful, and the flow was set ablaze, as is routine. The line of sight from the shack to the rig had between them that beautiful, powerful, impressive proof of the discovery of Prudhoe Bay field.

Oil was flowing at a daily rate of more than 1,100 barrels; it was burning because there wasn't ample storage for the volume of oil, nor an existing means of disposal.

That same afternoon the Dallas group flew to Denver for another meeting. Julius was constantly on the phone, attempting to stay current with the progress of the drill stem test at Prudhoe Bay. He'd call Anchorage, then Dallas, where senior staff member Stuart Mut nervously replaced Louis in his absence. Julius knew exactly what he was doing. He was working his way up the company's power ladder, making sure top executives got the information from him, rather than outside sources. At one time or another, he called the corporate office in Philadelphia and ARCO's Chairman Robert O. Anderson, who was by then visiting in Palm Springs, California. Eventually he even tracked down President Thornton Bradshaw, on the yacht *Sunbeam*, in Palm Beach, Florida.

Because of the bogus report of a fire, the company had to report something to the media. All through the late afternoon and early evening, Julius was on the telephone with management, trying to agree on a press release. This was no simple matter. The release had to satisfy a variety of interests: the Securities and Exchange Commission, because ARCO management didn't want to be accused of manipulating the stock; the general public, who doesn't understand the oil industry; and the rest of the oil industry, whose members would be furiously reading between the lines. Most importantly, ARCO was interested in treating this astounding news in a way that would keep secure, significant proprietary information.

Back then, the time differential between Denver and Anchorage was four hours; Florida was six hours later. The next morning, at 6:00 a.m. Florida time, Julius wrapped up the last call to President Bradshaw and reported the events that had unfolded during the previous eighteen hours at Prudhoe Bay. Then he took a sweet two-hour nap, packed his mukluks, and headed for Dallas.

After the dramatic events of March 12, three more open-hole drill stem tests were made; all three recovered formation water, meaning we found the bottom of the oil zone. The well was drilled deeper to economic basement, that is, below which there is no hope of success. Pipe was placed in the well in early April, so production testing could begin.

In May 1968, two Lisburne zones were tested. These were production tests through perforations in the pipe that had been set in the hole. Production tests are safer and more definitive because of being inside the pipe. The engineers had at hand everything they had learned during drilling, which enabled them to precisely test the well in the best zones. One test was essentially a retest of the previous open-hole drill stem test. Another production test was in a previously untested zone. Both zones produced oil, but not in large amounts — one at 493 and one at 434 barrels of oil per day. These were not through a choke or restriction in the flow lines. They were open flow or unrestricted tests and not too important. The results were frustrating but not surprising, knowing the general history of carbonate (limestone) reservoirs.

Enthusiasm for the Lisburne as the primary predrilling estimate was somewhat validated when future wells in the general Prudhoe area produced from the Lisburne. They were economically viable solely because production from the Sadlerochit justified the costly infrastructure of gigantic production facilities and the pipeline to move the oil to market.

In 1991 the U.S. Department of Energy reported, "The Lisburne reservoir, located directly below the Prudhoe Bay Sadlerochit formation and operated by ARCO as the Lisburne Participating Area of the Prudhoe Bay Unit, had original oil-in-place of about 3.1 billion barrels of oil. Recoverable reserves are estimated at 1.5 billion barrels of oil."

As of January 1, 2001, the Lisburne had produced 135.1 million barrels of oil. This would indicate that the recovery of oil-in-place was less than had been expected. Limestone reservoirs characteristically do not have great recovery factors. Daily production on the same date was just over 10,000 barrels per day. This is a significant amount of oil, but a drop in the bucket of U.S. daily consumption. Based on the production to date and the current rate, the reserves probably have been drastically reduced. Lisburne production at Prudhoe Bay appears to be a footnote at best. Fortunately, ARCO and Humble had the future testing of the Sadlerochit formation to blunt the disappointment of the Lisburne. They also had high hopes for the Prudhoe Bay *Sag River State* No.1, the confirmation well.

During the final production testing of the *Prudhoe Bay State* No.1, ARCO and Humble had started drilling the Prudhoe Bay *Sag River State* No.1. This well was designed to help the geologists, geophysicists and engineers determine the limits of the field and the all important oil–water contact. The depth of the oil–water contact below sea level was the key to the extent of the Prudhoe Bay oil field. So, the big question was, just what would the Sag River reveal?

Chapter 15

Confirmation!

The news from the *Prudhoe Bay State* No.1, as good as it was, would only be of academic interest if we could not establish a larger potential producing area than had been found. The confirmation of the discovery involved four main issues. The most important matter was to gain some sense of the size of the oil field. The location of the oil–water contact beneath the oil would yield that answer. There was no other way to establish the oil–water contact but to drill for it. Another point of control for the geological and geophysics of the area would give us more insight in to the shape of the structure. Before *Prudhoe Bay State* No.1 was drilled, Rudy Berlin and Marv Mangus made a projection based on their knowledge of the geology and geophysics. It could not have been much more precise. There was no reason to think a different location would yield surprises, but the test would yield a velocity survey to go with the one from the *Prudhoe Bay State* No.1 to yield a better understanding of the seismic velocity. Another issue was that the Sadlerochit sandstone and conglomerate rocks in the discovery were nothing like the Sadlerochit our surface geologists had seen on outcrops. We needed to see if these spectacular reservoir rocks covered a wider area. This was before the geoscience studies revealed this was a buried delta system. And finally, it was urgent that we find a drilling location for the test on which we could operate through the summer months. This required thousands of cubic yards of gravel for construction of the well site and an airstrip to service the operation. Of all places doing oil exploration, in Arctic Alaska the adage “time is money” was a truism. If this oil field was to be a worthwhile venture it would unleash an astounding amount of drilling and construction activity. At some place and time we would need a year-round base of operations. Another temporary set up like that used at the discovery was impractical for the job at hand. These goals can be summarized as establishing the size, shape and continuity of resevoir and operation feasibility of the discovery area.

It was vital to learn answers to the reservoir questions, but the gravel situation trumped them because there were any number of locations we could have picked to get the reservoir answers. Gravel was important because the drilling operation would continue through the summer, so we needed a well location that was accessible by large aircraft in the event of a drilling or safety crisis. We needed six feet of gravel under everything to prevent our activities from thawing the permafrost and making a quagmire. Two items went into the gravel equation — availability and accessibility. This combination occurred at the Sagavanirktok River about seven miles southeast of the *Prudhoe Bay State* discovery. The terrain was cooperative; it was relatively flat away from the banks of the Sag River.

Historically after a discovery, development occurs on a location-by-location basis and usually in distances less than three-quarters of a mile. As noted, the location chosen on the Sagavanirktok (the Sag) was seven miles southeast of the discovery. This was a bold step out. It was also bold because the sub-sea level was about 400 feet lower than the top of the gas–oil contact in the discovery. If this location would have been in the water table below the oil, it is conceivable that there could still be oil in the intervening area between here and the discovery. That would have been a real teaser because it might not have been sufficient to build the producing infrastructure and the pipeline. Drilling the confirmation well was a make-or-break situation.

There was no sense of solemnity; to the contrary, the decision was made with alacrity. Julius Babisak put it this way, "The location of the second well was an exercise that was not too scientific. We had our location, and there was no real reason to anguish over it." Louis Davis was a great man, but the uncertainties inherent in exploration made him extremely cautious and reserved. Of all things, he asked to see the seismic map, which he pored over, and in time demanded to see the exact shot point of the well location. Louis's insistence on a particular shot point seemed uncharacteristically equivocal. Not to worry; Julius again:

> He [Louis] was the most pragmatic man I ever knew. Seven miles didn't bother him because he knew for things to go well [with the future of Prudhoe Bay], the field had to extend that far, but a dry hole would be fatal.

ARCO's vice president for exploration worldwide and Louis's boss, Mr. Hamm, had the final vote and never batted an eye in confirmation. "If a prospect is so weak a second location is a problem, it is no good. Never make a seismic point be that critical."

Marv Mangus and Wray Walker again staked the well location. This was a formality and requirement for a drilling permit, but any location in the general area seven miles southeast of Prudhoe and close to the Sag River would have done (Figure 14).

The drilling department located Nabors Drilling Company from Calgary, Alberta that had a suitable rig stacked (stored) roughly seventy miles to the west. It was moved north to the near-shore sea ice, east on the ice, and then south to the location.

Enter John Rousi and Joe Dunn, two "magicians" who played a vital role in the success at Prudhoe Bay for years, largely because they knew how to make things happen. John Rousi was a warehouse man, finder of material, transportation coordinator and expediter *par excellence.* He was hundreds of miles behind the lines in Fairbanks, but he knew the ropes. The drilling crew on the front lines knew John was unfailingly dependable in meeting their needs, and on time. In a remote operation it is critical to get it right the first time, and John did.

Joe Dunn, in construction, was an expert in moving heavy equipment from one place to another and building earthen structures. A man of great physical stature, yet quiet, unassuming and effective, it was Joe's responsibility to move the gravel. He made sure that the airstrip, location and camp pads were quickly and safely built, thereby accommodating all activities during the summer thaw.

Everybody moved so expeditiously that the *Sag River* well, and the final testing of the *Prudhoe Bay State,* were operating nearly contemporaneously. It was a fine example of teamwork. Because of all this activity, rumors were rampant. The *Anchorage Times* newspaper in the early summer of 1968 reported, "Were ARCO not so greedy and protective of information derived from *Prudhoe Bay State* No.1, the company could have signed an agreement with British Petroleum to earn some BP acreage." Julius Babisak was in on what really took place.

> Three of us from ARCO and three top people from Humble met with key people of British Petroleum, North America. We discussed the exploration picture. I had with me the closely guarded electric log of the well, showing the gas column, the oil portion, and the sands full of tar. We had much common data between us, but somewhat different ideas of what it meant. We offered a generous work program to earn part of their acreage. They were impressed and receptive. We surmised that their acreage on the northeast flank of the structure was in the area of the truncation and unlikely to have any oil sands present. They did not believe that a truncation existed [subsequent drilling proved us right] and assumed the map that everyone had was accurate and therefore [the

> northeast flank was] a rich site [lots of oil sand]. They insisted that they would not give up any of that, but would farm out their acreage on the southwest flank.
>
> We didn't want the northeast flank and gladly offered to embark on a vigorous drilling program on the southwest [flank, which ultimately was found to contain millions of barrels of oil]. BP in New York said yes, but needed approval from the highest management level in London, but they said no.
>
> Later, British Petroleum people told us that the London turn-down was more of a reluctance to accommodate a fierce foe [competitor Humble] than it was second guessing their North American office.

It would seem that corporate culture triumphed. Vengeance overcame a potential business relationship. Regardless of the motivation, BP made the right decision, as the future would prove.

The *Sag River State* No.1 was (started) on May 3, 1968 a month after the definitive oil Sadlerochit test in the *Prudhoe Bay State* No.1. It was drilled into the top of the Sadlerochit on June 17, 1968 and it had oil shows within a month of when the *Prudhoe Bay State* No.1 was tested in the Lisburne formation in May 1968. The impossible dream came true. The continuity of the Sadlerochit reservoir was so similar to the *Prudhoe Bay State* No.1 that even a derrick man could see it. On July 17, 1968, the *Sag River State* No.1 flowed oil at the rate of 2,110 barrels per day from only fifteen feet of the several hundred feet of oil "pay." Two subsequent tests, of ten and fourteen feet of oil sand had similar flow rates. The total interval of the three tests was 110 feet. The uniform flow rates were indicative of uniform porosity and permeability. The full force of the discovery at Prudhoe Bay began to be felt everywhere in the oil industry, the state of Alaska and the federal government.

The good indications from the *Prudhoe Bay State* No.1 had been thoroughly tested and evaluated. On June 24, 1968, it was plugged and abandoned. The well had served its purpose and would never be used as a producing well because permanent, producing wells had to be engineered differently to protect the permafrost and keep it frozen.

On July 18, 1968, ARCO Chairman R.O. Anderson and President Thornton Bradshaw released the data on the *Sag River State* No.1 to the press. Statistics cannot be defied because every event is part of some statistic. The *Sag River State* No.1 revealed things that before the fact would rightfully have been given little chance of happening. We had every reason to be confident of the structural location, but beyond that, the two unknowns that formed the justification for the well were both transcendent events. The water table that would set the limits of the field proved that the reservoir was full to the spill point. In many oil fields, the traps are not full. The other important factor established by the *Sag River State* No.1 was finding more oil reservoir than we had in the discovery.

ARCO–Humble drilled into the mother lode of reservoir rock contained in a fan delta deposit. It is true we didn't find this out until control from many more wells made it possible for people like Dave Hite and Kirby Bay to decipher it. The *Sag River State* did tell us there was no diminution of the reservoir in quality, and the icing on the cake was that it was "thicker." Finding the fan delta remains the peerless irony of the Prudhoe Bay story. The statistical chance for this particular formation to occur in conjunction with the Prudhoe structure boggles the mind. Many oil fields have been explored for one reason, but oil found for reasons unrelated to the original justification. At Prudhoe we drilled for the Lisburne as the objective, but the bonanza was in the Sadlerochit. We were in awe of what had materialized. Twelve years later Harry Jamison still couldn't believe it.

> It's still amazing to me that we could have moved out that far and done these things correctly and verified our conclusions with that well.

Just how dramatic and important to national oil reserves was the Prudhoe Bay discovery? Early original estimates of reserves were 9.6 billion barrels of oil. This estimate was exceeded by one billion barrels by May 1, 2007, when Prudhoe had produced 11.6 billion barrels. This was twice as large as the number two U.S. oil producer, the East Texas field with a cumulative total of 5.3 billion barrels and four times larger than the Wilmington field in California with 2.5 billion. Prudhoe Bay is larger than the combined output of the seven largest oil fields in the Lower 48 states, which have produced 16.5 billion barrels of oil.

Various predictions are that Prudhoe Bay cumulative production will reach 13 billion barrels by the end of its life span. These remaining reserves will be produced much more slowly than the oil produced until now. That future date of Prudhoe's demise is subject to debate. A ballpark estimate is in a range between 2010 and 2020. There are many variables; the most important is the magnitude of new discoveries, and especially the opening of Arctic National Wildlife Range to oil exploration. If these future additions are large enough to keep the infrastructure of the Trans-Alaska Pipeline operating, the Prudhoe Bay economic limit would be lengthened accordingly.

Without the existence of Prudhoe Bay field infrastructure, the additional 500,000 barrels per day from other fields in northern Alaska would never have been realized. The million barrels per day from all Alaska oil production was about 17 percent of U.S. domestic production in 2003. Prudhoe has alleviated the effects of two international oil shortages and continues daily to make a large contribution to the national energy needs.

There is so much oil in place in the greater Prudhoe Bay area, it remains a constant challenge for the petroleum engineers to discover ways to produce those reserves. The problem is that they are very viscous and are difficult to make flow to the wells. BP announced in their house publication *Horizon* in April 2005, that in-place reserves in the Ugnu area are estimated at about 12 billion barrels. Another irony, this is on the Colville structure that was mentioned in the earlier part of this story as being unproductive. It still is, unless they can find a way to release this shallow accumulation from the rocks. In the same area, the Schrader Bluff accumulation of oil in place is estimated at 15 billion barrels of oil. In another producing field in the area, Kuparuk light oil reservoir in the Milne Point field, they have developed a new enhanced recovery technique using a different combination of altering the injection of gas and water to increase recovery from the area by 100 million barrels and at an overall saving of $900 million for the field. The engineers think this technique may be applicable to those two voluminous accumulations mentioned earlier. More hope for the future.

Again and again it has been said that we were the possessors of blind luck. I adamantly disagree. There is no doubt that good fortune chased us across the Arctic. But we also did sound fundamental preparation so that the good fortune was divulged. The ARCO explorationists had identified: 1) what appeared to be a good trap; 2) shows on the surface outcrops and in some NPR4 wells; 3) potential reservoirs; 4) strong convictions, which we followed, against long-shot odds.

When one perseveres and follows the rules of the game, lightning is bound to strike. Seldom does it strike with the ferocity of a Prudhoe Bay, but strike it will, when, as at Prudhoe, the basics have been meticulously attended to. Here are Harry's comments again.

> People have said to me it was blind luck. I say that is wrong, and I have heard that so many times that it makes me want to throw up. As I said, there was a long series of events that took place leading up to this discovery and this confirmation. There was good technical work and it resulted in good fortune, for the company and everybody associated with it. The serendipity that occurs in our business is a vital factor. But if you don't afford

Field	Original reserves Millions of barrels	Cumulative Production 000's Bbls Production	Daily Production barrels per day April 2007
Badami	120	5,100	1,024
Colville	429	224,248	129,555
Endicott	1,000	451,249	13,918
Kuparuk	2,250	2,217,692	164,279
Milne Point	464	252,918	32,012
North Star	145	113,131	45,685
Prudhoe Bay	13,000	11,675,462	342,796
Point McIntyre			729,269
Natural Gas Liquids		545,146	64,495
Total	17,708	15,484,946	793,746

Field with discovery year	Location	Total production from discovery date until 1/1/98.
Prudhoe Bay, 1968	Alaska	11,675,462,000 (5/1/07)
East Texas, 1930	Texas	5,271,381,000
Wilmington 1932	California	2,497,262,000
Midway-Sunset, 1894	California	2,360,055,000
Kuparuk	Alaska	2,217,692,513 (5/1/07)
Wasson, 1936	Texas, West	1,964,141,000
Kern River, 1898	California	1,618,760,000
Panhandle, 1921	Texas	1,475,767,000
Yates, 1926	Texas, West	1,356,104,000

yourself the opportunity in the first place, by good thorough technical work and acquisition of land, you won't ever get there.

Blind luck or blunder factor, Harry Jamison dignified it with a lively word — serendipity. Much oil was found in the early days of oil exploration simply by poking holes in the earth. Later, promoters and drillers associated surface topographic features with oil fields; this led to the application of geologic principles, then geophysical techniques. One can apply all the right oil exploration techniques and fail to find oil. That has happened countless times. It has been said that oil can be found without science, but not without luck. Louis Davis, ultimately vice chairman of ARCO, in his inimitable serious way once looked me straight in the eye and said, "I would rather have a lucky geologist work for me than a smart one." I was never sure if or how he attached that to me, but I do know this, many of us were in Alaska by happenstance. I also know that for whatever reasons we were there, we were unbelievably fortunate to have been associated in so singular an adventure.

Those people in Richfield and Humble who, in retrospect, took what appears to have been only a small gamble by buying the leases, provided the indispensable building block. Retrospect is not a fair position from which to look, because that view makes it appear much too easy, for indeed they were taking steps of major proportions. It is incredible that they should have been so prescient.

The geologists in the story that were "going to hell" thinking there might be oil there must have done a quick retreat when they heard about the Prudhoe Bay discovery. Once an oil province has been established, everybody tries to get in the act. That didn't exactly happen after the discovery because the lease situation was stable, but those companies who had those leases watched events unfold. Oil people usually refer to an area where there is lots of production as being "oily," and the general Prudhoe area was that.

Early in the development of Prudhoe, oil was soon discovered

Figure 29: The North Slope Oil-producing Intervals.

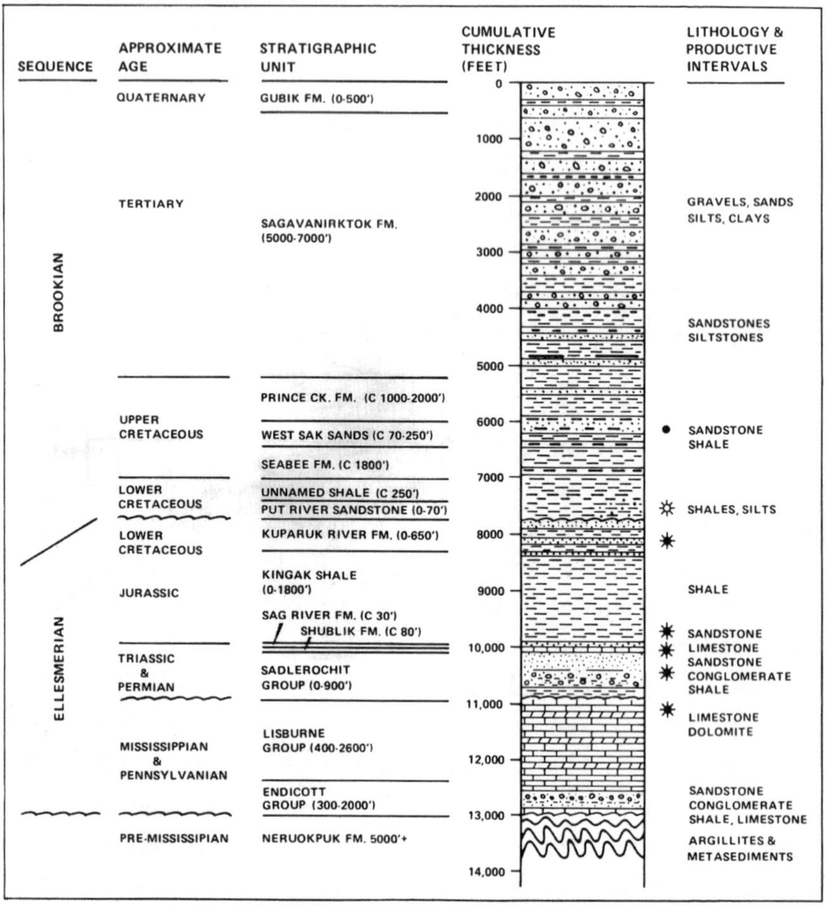

This summary stratigraphic chart compiles all of the oil- and gas-producing zones that had been found during the ten years following the discovery.

After Jamison, H.C. et al., Giant Oil and Gas Fields of the Decade 1968–1978. *AAPG (Copyright) 1980 reprinted by permission of the AAPG whose permission is required for further use.*

in the Kuparuk sandstone of Cretaceous age. This was the section that had all the oil in during the drilling of the discovery, but did not have good reservoir rocks. Geologists sometimes become overly exuberant when they see a little oil. Reservoir engineers like things certain and tidy, and were convinced for a while that it was uneconomic to be of any value. On January 15, 2004, the Kuparuk produced the two billionth barrel. So much for economics. The Kuparuk Oil field is the fifth largest in the United States and by the end of its life may vie for second.

The table on page 268 shows the Alaska fields with their estimated original reserves, the cumulative production as of May 1 2007, and the average daily production during April 2007. When the Natural Gas liquids are added to the average daily production, the field is producing 793,746 barrels of liquid petroleum a day. Natural gas liquids are light hydrocarbons similar to gasoline that are removed from natural gas before it is reinjected into the reservoir. There are seven oil fields in the Prudhoe Bay area on the North Slope of Alaska. As good as most of the six besides Prudhoe are, they all exist because of Prudhoe Bay. All of them together would not be viable without Prudhoe. A couple more fields were originally on the list and were unsuccessful. In addition to Prudhoe, there is one national-class field; one is approaching a giant category, and two into the major category. That is some record for an area that was damaged so badly by eleven dry holes that could have easily dissuaded ARCO to pass on the one that was successful. At this stage it would appear that all the best spots have been pretty well explored, but, who knows? Some enterprising geologist in the image of the famous Wallat Pratt may break the belief that all oil has been found, and find still another pool of oil.

Prudhoe Bay generated far-reaching dramatic events. The consequences rippled, no, surged, at times like a stormy sea. A true appreciation of the impact of Prudhoe Bay can come by learning what these consequences were.

Chapter 16

The Consequences

ARCO had already started to move ahead on two fronts before Anderson and Bradshaw had made their announcement about what had been found. There was an oil field to make ready for production, and there was a pipeline needed to take the oil to market. ARCO, Humble and British Petroleum began designing a pipeline in June 1968. These companies announced that they would build a 48-inch pipeline 800 miles to Valdez at a cost of $900 million. In August of 1969, the Department of Interior issued a permit for TAPS, the Trans-Alaska Pipeline System, to build the first section of a construction road. Four momentous things happened in 1970 that had the effect of scuttling the original design of the pipeline.

The first was The National Environment Policy Act, which became law on January 1, 1969. Being new, there wasn't much known about the Act's impact and implications. It wasn't long before TAPS learned that the impact would be absolute.

The next thing that happened was that two Native groups sued TAPS to withdraw waivers they had signed to let TAPS cross their land. A month later, five other Native groups brought a suit in Washington to enjoin Secretary of Interior Hickel from granting permits over their land. Thus began the protracted negotiations that resulted in the Alaska Native Claims Settlement Act.

The third distraction was that secretaries of interior had been granting unauthorized rights-of-way extensions for years under the Minerals Leasing Act. Environmental groups caught what had become customary, but nevertheless an overstepping of authority, and this required an act of congress to fix. To fix this, congress passed, and on November 16, 1973 President Nixon signed, The Trans-Alaska Pipeline Authorization Act.

Finally, TAPS had done soil studies that waived the red flag of permafrost, and its effect on a buried pipeline. The Interior Department involved the USGS, which also began its own investigation of the project. Permafrost and the burial of the pipeline became a major challenge to both the industry and government scientists and engineers.

Dealing with these matters seemed endless. Environmental groups fought it tooth and nail. Some of their concerns were well taken; some were off the wall. Geoff Larminie, BP's manager at the time, answered critics at one of the many hearings in Anchorage by opining, "Indeed it all too often seems that these many critics who so freely use the term 'ecology' know as little about ecology as they know about Alaska." He asked rhetorically, "In other words, why should we have to continually account for our knowledge, and they never have to account for their ignorance?"

As part of the National Environmental Policy Act, there was a requirement to list and discuss alternatives to the proposed project. Humble beefed up a tanker and succeeded in getting it to Prudhoe through the Arctic Ocean via the Northwest Passage. It returned with a gold-painted ceremonial barrel of Prudhoe Bay crude oil. There was no shortage of other ideas from outside the industry. General Dynamics proposed a nuclear submarine tanker fleet of 170,000-ton tankers to take the oil to Greenland for trans-shipment. Dr. Edward Teller, the nuclear scientist, proposed building a North Slope harbor using nuclear explosions. Extending the Alaska railroad would have been able to do the job with sixty-three trains a day, each way.

Monorail, trucks, planes, blimps and even a pipeline large enough to carry tractor trains carrying the oil in barrels were suggested.

The United States never had a formal treaty agreement with the Natives, which would have precluded protracted legal wrangling over the ownership of the land. Therefore the Alyeska Pipeline Service Company, owned by seven of the potential producing companies, became the catalyst for a settlement that the Natives had been trying to obtain for many years. Alyeska Pipeline Service Company could not proceed without this festering issue being decided. Without a settlement there was great potential for litigation because the pipeline may have crossed lands claimed by different Native villages. There are three different Native ethnic groups in Alaska: the Inuit (formerly referred to as Eskimos), Indians, and Aleuts.

It was Walter J. Hickel who came to the rescue of the Native population. Hickel, who a few years before had resigned as governor, went to Washington, DC to serve as secretary of the interior for Richard Nixon. Secretary Hickel played a vital role in getting the Alaska Native Claims Settlement Act passed. Although the other cabinet officers were aligned against the terms that Hickel proposed, Nixon said, "Let's hear from Wally." Hickel wrote in his book *The Alaska Solution,*

> Mr. President, I agree with my [Alaskan] colleagues; there is no legal claim here, nor is there an economic claim. Mr. President, it's a moral claim." As quickly as that, Nixon's face lit up. I knew that expression well. This was the positive Nixon who enjoyed doing things that were innovative and right, the side of him that, tragically, was lost later. He dismissed the arguments of the other Cabinet Officers and said, "I'm going with Wally.
>
> From that day on, Nixon was the strongest ally of Alaska's Natives in Washington…without Nixon's support…a settlement…would never have happened. Nixon [later] met

> with Don Wright, president of the Alaska Federation of Natives, and after hearing Wright's impassioned plea, Nixon committed to veto any bill that the Natives felt didn't meet their needs.

On December 14, 1971, the Alaska Native Claims Settlement Act passed Congress, ceding 40 million acres of land plus 962.5 million dollars to thirteen regional Native corporations and many village corporations. Before he would sign the bill, President Nixon wanted to know if the settlement was agreeable to the Native community. So the AF voted 511 to 56 to accept the terms, in which they would receive title to one ninth of Alaska and nearly $1 billion. (One example of what this means to the Native corporations: the North Slope Borough, headquartered in Barrow, Alaska, received $193 million dollars in 2001 from oil tax income from the Prudhoe Bay oil production.)

The Native Claims Settlement was the step the industry needed to proceed with construction. Much earlier, Ralph Cox, as ARCO's Alaska manager, asked me to accompany Richard G. Dulaney, president of Atlantic Pipeline Company, to Prudhoe in order to show him our operation. That was during the late evaluation stages of the discovery and confirmation wells, and the pipeline was an obvious subject for discussion. On the way back to Anchorage, Dick pulled an envelope from his pocket and began to surmise what a pipeline might cost. It was always my belief that the figure was not meant to be anymore than a point of departure for more investigation. Although he used the $900 million figure, there were so many unknowns at that point that Dick probably never dreamed his offhand estimate would be picked up and bandied about as it was. But lo and behold, a year later the first pipeline plans from ARCO and Humble estimated the cost at $900,000,000.

The Trans-Alaska Pipeline operated by the Alyeska Pipeline Service Company is defined in superlatives. Robert O. Anderson, ARCO board chairman, described it as the greatest man-made project in the world. James P. Roscow, in *800 Miles to Valdez*, on the building of the Alaska pipeline says it "dwarfed any other modern-day industrial project." He likened it in scope to the building of the transcontinental railroad and the Panama Canal. TAPS would not have been possible with the technology back then. Dr. Hal Peyton was a structural and civil engineer at the University of Alaska, and then later worked on the pipeline. He was widely recognized as an expert in cold-climate engineering for twenty-three years. He said,

> It's been a first of a kind project in every sense — trying to build a project in the far north, with its technical and construction problems, and the complexity of trying to convey to people…how to design and construct there…The utter sheer size of the project…until you've lived with it a long time, you readily don't begin to comprehend how massive it is.

After all the permits and approvals had been received, Alyeska began the haul road from which the pipeline would be built. ARCO and Humble Oil ordered the pipe in early 1969 and the first delivery to Valdez arrived in September 1969. The $100 million cost of the pipe was the least expensive outlay of the pipeline, about 1.25 percent. When the road was completed, Alyeska began stringing the pipe together on March 27, 1975. The 800-mile route begins at Prudhoe Bay, goes through three mountain ranges, and ends at the Port of Valdez. The highest pass is Atigun in the Brooks Range at an elevation of 4,790 feet. The pipe is buried for 420 miles of the route where the ground was deemed to be stable. The remaining 380 miles are above ground, 80 percent of which is supported by vertical support members (VSMs). These are made of eighteen-inch steel pipes welded in an "H" shape and buried upright in the ground. They are used in permafrost areas to isolate the heated pipeline from the permafrost. Each vertical support member is fitted with a passive refrigeration unit to keep it from thawing the permafrost in which it is buried. The pipe is fastened at about 600- and 1800-foot intervals. In between these secure points it is mounted on the VSMs so it can slide from side to side or along its length, in the event of an earthquake. The VSMs are also equipped with motion detectors to monitor movement from any source. These above-ground pipeline sections are covered with fiberglass insulation that is three and a half inches thick. The insulation is encased in galvanized steel secured by steel bands.

The line is built in a trapezoidal manner that is in zigzag shape, with the "zigs" extending diagonally from one straight section to the next. The alternate straight sections are 300 feet long and diagonals are 180 feet long. It was built this way to compensate for the great temperature change when the hot oil went into the line. This temperature at the wellhead measured 160°F. It stabilized in the pipeline at 140°F.

It is now 2007 and the last few years the pump stations have been upgraded and made more efficient to handle the present oil through-put, which is about 750,000 barrels per day. Maximum through-put early in its life was over two million barrels per day. Originally, three 35,000-horsepower aircraft-type turbine engines with pumps in each of twelve pump stations were installed to push the oil along the pipeline. They started the operation slowly at the rate of 300,000 barrels per day, then went up to 600,000 in another thirty days. It took the first oil thirty-one days to reach Valdez. At first the oil was traveling only about one mile per hour, and because it was going into the cold pipeline it was similar to asphalt, but the hydraulics were still within maximum design pressures. There were a million electronic parts in the leak detection system, and they didn't want to rely on the detection system entirely. There was a crew walking with the flow to make sure all was going right, another crew followed in twenty hours, and another one in another twenty-four hours. The first oil arrived in Valdez at 45°F and gradually stabilized at 140°F. Unfortunately, disaster struck at

Pump Station Eight when extreme low temperatures cracked a pipe bend, and a fire killed one worker and injured five others. This station was bypassed, so the accident did not delay the process.

The Valdez Terminal has eighteen 510,000-barrel storage tanks. At a million barrels per day this was storage for nine days of production in case it was necessary to adjust to tanker schedules. The ARCO *Juneau* sailed with the first load of Prudhoe Bay crude oil on August 1, 1977. Sometime in 2003, the 14 billionth barrel entered the pipeline.

During the building of the pipeline there were 70,000 workers involved, with the highest peak of 21,800 on the job. A major problem occurred with the welding inspection. The contractor got behind with inspections and resorted to falsifying records. There was no way of knowing if the welding was faulty, short of reviewing 30,800 of 65,000 welds. Alyeska did the job itself and found 3,955 welds about which there might be questions. Alyeska eventually rewelded thirty-one, but twenty-one of these were actually passable. The government inspectors granted waivers on three of those because they were under a river. Alyeska estimated the cost of this boondoggle at $55,000,000.

The nearly four years during the permitting process were put to good use in the line's design. One of the engineers said that it was designed foot by foot. This may have been hyperbole, but certainly every foot had to be planned. Studying, planning for and negotiating the permafrost terrain were all-consuming tasks. Much of the line was also built on land with high probability of earthquakes.

After twenty-nine years of use, the pipeline design has acquitted itself well. The legal profession has a term for it, *res ipsa loquitor* (it speaks for itself). In 2001 a study indicated pipeline integrity has been stable and it can physically operate conservatively if the through-put is still at 500,000 barrels per day in 2034. In 2001, the reliability of the pipeline was 99.6 percent. Six major leaks (greater than 1,000 barrels) have totaled 31,600 barrels, half of which was due to sabotage. The frequency of spills is now a third of what it was twenty-five years ago. The pipeline is well maintained; one example of the annual maintenance is the massive expenditures of over $25 million spent each year to find and control corrosion. British Petroleum reported in February 2006 that the 15 billionth barrel of oil had passed through the Alaska Pipeline in December 2005. It currently transports less than a million barrels per day, which still is 17 percent of the national crude oil production from this one golden egg of a field.

In 1964 the largest recorded earthquake in North America, with a magnitude of 9.2 on the Richter scale, occurred in Prince William Sound, forty miles west of the Valdez tanker terminal. The pipeline also went through a zone known for its potential for earthquakes. The design for the earthquake contingency thus was every bit as vital as the preparation for the permafrost. The fruits of the earthquake design paid off when the Denali Fault came alive with a 7.9 Richter magnitude earthquake on November 3, 2002. This was the strongest strike-slip earthquake since the San Francisco earthquake of 1906. It was also significantly more powerful than the San Francisco event. There was not much in the news about the 2002 shaker because there was so little devastated habitation to write about. A fault is a large, steep crack in the earth. A strike-slip fault means that if an earthquake occurs, the rocks on the opposite side of the fault from your position will slide to your right or left depending on the dynamics of the earthquake.

The fault passed beneath the pipeline with a right lateral offset. Think of a straight line drawn perpendicular across the fault before the earthquake. After the earthquake that line would be broken by the movement on the fault and the ends would be about twenty-nine feet apart. This was the maximum displacement of the Denali Fault. The movement lasted ninety seconds. So, if you had your feet on opposite sides of the fault, it might not have been too difficult walking one leg on the opposite side of the fault as it moved past you. Right lateral offset means the end of the broken line moved to your right.

William Hall, et al. reported on this with a paper titled, "Performance of the Trans-Alaska Pipeline in the November 3, 2002 Denali Fault Earthquake." The authors found that the earthquake was a real-time test of the design, which was essentially equal to the force of the quake. The important discovery was that the pipeline did not break. In the vicinity of the pipeline, the right lateral displacement was about eighteen feet. There was also about two feet of vertical displacement. This means that if you crossed the fault moving north, you would have to negotiate a two-foot step up. The sliding mechanism on the vertical support members worked as planned. The way the land shifted, and the alignment of the pipeline relative to the fault, caused the pipeline to become compressed. This compression was about eleven feet and the displacement was taken up in the bending and bowing of the pipeline, as planned. The vertical support members were more severely damaged than expected and studies are investigating remedies. The authors could find no other examples of a large diameter pipeline that has been designed for, or tested by, an earthquake.

The economic limit of the pipeline will be reached when the available oil to pump through it approaches somewhere in the range of 200,000 barrels per day. Rest assured, the Trans-Alaska pipeline will not be shut down until every alternative is considered. Indeed, 200,000 barrels is a lot of oil — enough to boggle the mind. In the South 48 that amount of oil would be a bonanza for the owners and the nation. Not a single existing oil field in the United States produces that much oil per day. There is, however, a plant in Canada's tar sands that is producing just over 950,000 barrels of oil per day. The tar sand reserves are over 113 billion barrels.

For readers turning these pages, the time from discovery in mid 1968 to the beginning of production in 1977, is fairly rapid. In reality, though, for the project's proponents it seemed interminably slow for the Alaska Native Claims, the pipeline engineering and permitting, to grind through the process. The delay in building

TAPS was disappointing, but ARCO and Humble had plenty to keep them busy. ARCO, as operator, designed and built a comfortable camp and began preparing for the day when the pipeline would be ready to take oil. As part of the permanent camp complex, one of the first things ARCO did was to build a topping plant, or a small refinery, to make various grades of diesel and gasoline. Fuel had been a very large item for transportation and storage. From then on it was one of the least expensive commodities in the operation.

Prudhoe Bay field has a gas cap, that is, a gas reservoir above the oil. The expansion of the gas cap provides energy to make the oil flow. Water is under the oil, so the oil had some amount of energy from the water pushing from below. Most of the gas produced with the oil is reinjected to help maintain the gas cap pressure. Water is injected below the oil, so both the injected water and gas have helped to recover the maximum amount of oil from the reservoir. When the oil is all produced there will remain a very large gas reserve for which almost everyone in government, those who own the gas, and anybody who thinks they have a chance to make a buck, is looking for ways to market.

Three other large, necessary operations unfolded from the beginning days and pretty much ran up to the time when the pipeline was ready to accept the oil production. The first was rather mundane in implementation, but it was pervasive. Joe Dunn and the construction people had long since learned that to the way to keep the permafrost frozen and stable during the summer thaw was to cover it with five to six feet of gravel. This was done during winter to minimize damage to the tundra and underlying permafrost. There was a continual building of roads, drilling pads and pads for oil flow lines.

The first oil field operations were to begin drilling the producing wells. Protecting the permafrost was the primary concern. At first the drilling pads for six wells was 240 feet by 360, with 120 feet between wells. Some pads contained eight wells and were correspondingly larger. The five- to six-foot gravel pad protected the permafrost from surface exposure; there was exposure under the gravel from the heated drilling fluid as it circulated in the wells. It took some time to learn to drill the first 2,200 feet of hole where there was permafrost. The technique was to drill this section as fast as possible to minimize the exposure to the permafrost by the drilling mud. At first this was sometimes four weeks. By 1970, with better rigs and experimentation, they reduced it generally to less than four days. This part of the hole was cased within $13^{3}/_{8}$-inch pipe. After the production pipe of $9^{5}/_{8}$-inch was installed, the annulus between the two sets of pipe was filled with a thick oil-based mud. This mud was a way of eliminating water from the annular space to prevent freeze-back expansion. The thick mud also was also an insulator to reduce thawing of the permafrost.

After production started, the drilling and production engineers found that thawing induced by production was not as extensive as they first calculated. This and smaller more efficient drilling rigs were chosen so the derricks could be laid down in less space during rigup and teardown.

Eventually there were forty drilling pads. One well was drilled vertically, and all the rest were drilled directionally to a different predetermined regular spacing pattern at the producing depth. A specific pattern was based on engineering studies that postulated the most efficient drainage between the wells as they were produced.

Originally the reservoir engineers thought something like forty or fifty high-rate producing wells would meet the production requirements. With additional studies, and in practice, they found that a phenomenon called coning occurred — by gas, where the wells were near the gas cap, and by water near the water table. This is a complicated subject, but in simple terms a cone of gas or water forms that inhibits the most effective draining of the reservoir. There are many factors involved, but too-rapid production is the cause of the problem. When this happens there is no going back; the cone remains.

The Prudhoe Bay field began producing in 1977, and by 1979

was producing at Alaska's authorized rate of 1.5 million barrels per day from 200 wells. That is an average of 7,500 barrels per well per day. The operators sustained that production for nine years, but only by increasing the number of wells. This means that the decline began almost immediately. There is nothing unusual about that because all oil reserves are finite, and when any amount is produced, less remains. Decline is inevitable the instant production begins. At the end of nine years, in 1986, the number of wells had increased to 870 to maintain the 1.5-million barrels per day, but the average rate had become 1,724 barrels per well per day. This rate is calculated to show the average degree of decline over the nine-year period. In the real world, well production varied quite widely. Some wells averaged (during those peak years) 11,000 barrels per well per day.

Enhanced (previously called secondary) oil recovery was begun in 1982. Two different techniques are used: pressure maintenance by water drive, and gas injection. All but a small amount of the gas used for fuel is reinjected into the reservoir. In the early 1900s enhanced (secondary) recovery was not begun until most of the natural energy drive was exhausted. Later the engineers learned that it was more efficient to begin the enhanced recovery techniques early in the producing life of an oil field. In short, enhanced recovery helps recover oil that otherwise would stay in the ground, and at Prudhoe Bay it was begun fairly early. *The Oil & Gas Journal,* 1988, reported that an enhanced oil recovery survey reported 50,000 barrels of oil per day total production and 5,000 barrels per day enhanced production. The impact of enhanced oil recovery will increase as Prudhoe continues to decline in production.

In 1999 some of the oldest wells were some of the best producers. The operators drilled some wells horizontally into the producing rock formation. This exposes much more of the bore hole to the producing rocks and should improve production. Nineteen of these wells were compared to nineteen vertical wells, and the horizontal wells produced oil better than the verticals in the first couple of years; then the verticals produced better. Over the life of the wells it appears there may be little or no difference. The difference in producing rates between wells could be caused by local changes in lithology and permeability in the reservoir.

Drilling new wells to maintain production is fast losing its appeal because the rates are low and in 1999 about 20 percent of the wells drilled were completely dry. Also, many of the wells produce high volumes of water fairly early in their producing lives. Water percentages of 80 percent and even 95 percent are not uncommon.

The rate of decline for Prudhoe Bay is variously estimated between 4 and 12 percent per year. Those rates mean that the life of the field would end between 2010 and 2020.

The other large task facing the production engineers was the design, building, moving and installing the production facilities. What made this a formidable task was that because of its size no oil field had ever required this magnitude of equipment. The purpose of these facilities was to process the crude oil to make the crude pipeline ready. The production facilities removed the gas and water, which were reinjected to help maintain the energy in the reservoir. In most oil fields this equipment is made in modules as needed for one or several wells. It was determined that they should be fabricated in Tacoma on Washington State's Puget Sound, as field-wide modules to be barged to Prudhoe Bay and assembled. Every step of this operation was gargantuan.

The units were typically ten stories tall and as long as a football field. After leaving Tacoma, they were barged through what at times can be among the most treacherous seas in the world. The window for delivery at Prudhoe Bay was usually a couple of weeks in late August or early September, when the ice typically moves offshore. The barge captains would sometimes have to stand by for weeks at a time, waiting for offshore winds to open the waters. The hope, then, was that the winds would last until they could unload and get out. Good fortune must have been on their side, for they always made it both ways.

These huge buildings with their production piping and machinery were welded to the barges. On arrival, the barges were grounded as close to shore as possible. There were no docks, so gravel ramps were built up to the level of the barge deck. An ARCO base manager remembered their weight as being over 5,000 tons. Low-profile, specially built crawler tractors with hydraulic jacks transported the modules from the barges to waiting pilings for supporting the structures, where several would be connected together. Piling was chosen rather than gravel pads because the perpetual heat would eventually have degraded the permafrost. The piles were frozen in place and the modules were six to eight feet above the tundra, so there was no heat transfer. It was astounding to me that this was all accomplished without mishap.

We have mentioned natural gas only in passing. You may remember that the ARCO-Humble, *Prudhoe Bay State* No.1 discovery well drilled into the gas cap of the Prudhoe Bay oil accumulation. At that location there are over 400 feet of gas column and sixty-odd feet of oil column. The gas reserves in the greater Prudhoe Bay area are estimated at 37.5 trillion cubic feet. The interested parties are the oil companies who own the gas rights, the State of Alaska, the U.S. Government, and the Canadian government, should the gas line go through Canada. There have been many ideas floated about how to get the gas to market. Besides the four principals just mentioned, there have been many others who have offered their two-cents worth to the discussion. As far as we know there is no consensus on the subject, and probably will not be until the South 48 gas market runs short of fuel.

In 2007, BP's house magazine *Horizon* published some new and interesting information on the gas situation. Few outside the energy business have heard about gas hydrates. These are gas deposits that have been recognized in deep ocean bottoms many places on the globe. There has been considerable technical writing about them, but not much is known about how they might be produced.

The U.S. Department of Energy recently funded a $4.6 million test in the Milne Point oil field northwest of the Prudhoe Bay field to drill a well to 3,000 feet to study the gas hydrates in the area. The gas is in a frozen mixture of sandstone and ice. The operators had to drill the test well in a manner that would not thaw cores, and keep them frozen until the studies could be done. Gas hydrates occur right at the base of the permafrost and have been known in the area for years. They have not been studied because they were of no value, and in fact occasionally would cause drilling difficulty while drilling for the deeper oil zone. BP, the University of Alaska, Arctic Slope regional government and Doylon Drilling Co. also participated in the study.

Here is the bombshell. Government geologists estimate on the basis of this work that that there are 450 trillion cubic feet of gas in the hydrates. That is twelve times the conventional gas at Prudhoe Bay! It would appear if the diverse interests ever get a viable and agreeable plan, gas will represent an energy "second coming" for the State of Alaska.

In January 2002, The Alaska Oil and Gas Association released a noteworthy study. It calculated that the oil industry continues to spend $2.1 billion each year in Alaska. This is over and above the royalty that goes into the Alaska Permanent Fund. Payroll accounts for $422 million, and $1.7 billion goes toward goods and services. All this activity sustains 33,600 jobs, which generate another $1.4 billion, and the value added to the Alaskan economy benefits by another $1.8 billion. The industry's direct expenditures, combined with the economic activity it creates, totals $5.3 billion per year. As United States Congressman Everett Dirksen once said, "A couple billion here, a couple billion there, and first thing you know, you are talking about real money."

I was not able to find an accounting of the oil industry's total investment in Alaska, but it has to be an astronomically large figure. A colleague from those days ventured a guess of a minimum $30 billion, and likely much more, plus $8 billion for the pipeline.

Manufacturing, sea lifts and construction went on continuously, while field wells were drilled and pipeline political battles raged. Drilling expenditures were additional costs. Over 2,500 wells were drilled. At an estimation of $1 million each, they tallied $2.5 billion.

As mentioned before, the oil industry is still spending $2.1 billion per year in Alaska. If at this late stage they are spending that amount, it is safe to surmise they have been spending even more since 1970, which would add up to another $67 billion for ongoing expenses at the time of writing. A large portion of these expenditures would likely have been spent ten years before revenue commenced, and the cost of interest on the invested money would also be staggering—$100 billion might be close to the mark, plus or minus. Unfortunately, all these costs are obviously ambiguous, and no attempt was made to account for the changing value of a dollar.

The State of Alaska, on the other hand, has enjoyed tremendous income since production began at Prudhoe Bay. The $900 million from the 1969 State Lease Sale was the first really large lump sum awarded to the state. Substantial funds began to flow from taxes and fees during the many years of pipeline construction. Sizeable revenues surfaced when oil started arriving in Valdez for shipment to the South 48. This generated huge oil royalty payments to the state, which was the value of 12.5 percent on most leases, but 15 percent on some other leases.

In 1976, Alaskan citizenry approved a constitutional amendment creating the Alaska Permanent Fund. Its mission was to receive and invest 25 percent of the royalty income. Governor Hickel believes the intent was preempted by then-governor Jay Hammond. Here is what Governor Hickel wrote in his book *The Alaska Solution.*

> But Governor Hammond's goal was not to make additional funds available to the needs faced by future legislatures and governors. His motivation was to prevent government from

> using even the earnings from the fund, especially for the major capital projects that would encourage the development of Alaska's natural resources. Once voters approved the fund, he unveiled a device that would keep the money forever out of government's hands. He called it a dividend to be sent to every Alaskan every year. Voters would be seduced by their annual checks from the government and prevent any other use of the Fund's earnings. In essence, the Fund would absorb billions of oil dollars, but instead of putting the earnings to public use, Hammond created a dividend-paying machine.

Governor Hickel makes sense. The benefit of the oil dollars should accrue not only to today's Alaskans, but also to future Alaskans. The dividends paid now will be gone forever, but the money put into improvements would benefit future generations. Right or wrong, Governor Hammond knew that once the citizens of Alaska got an annual check, his idea would live forever. Can you imagine anyone getting elected to the state legislature whose platform was to take everyone's annual check away from them? I don't think so. So for now, if you live in Alaska, you get the dividend.

The Alaska Permanent Fund has paid the dividend for twenty five years. The amount given the first year in 1982 was $1,000 per citizen. The lowest dividend, $331, was issued in 1984. Since that time it has increased most years and peaked in the year 2000 at $1,963. The total paid out that year was $1.15 billion. A family of five received just under $10,000. The dividend decreased in 2001 to $1,850.28 and in 2005 to $845.76, and in 2006 it was $1106.96. Over the life of the fund through year 2000 it had earned $21.3 billion, 42 percent of which has been paid in dividends. The remainder has been retained and added to the principal, not only to increase the principal but also to inflation-proof the fund; however, recent figures indicate the need for some belt tightening.

When I was attending the many forums on whether or not the

pipeline should be built, one theme inevitably would emerge. It went something like this: "You will produce the oil, and take the money and leave behind a lot of junk, and a wrecked economy, just like the gold miners did."

This always seemed to me to be a defeatist argument, as well as specious. In part it was true, because the gold had no value in Alaska, so it had to be taken out to market. And, yes, much of the money would still have to be taken outside of Alaska because of limited investment vehicles there. It seemed even then that Alaskans should take some responsibility for preventing the take-the-money-and-run argument. It appears that, to date, Alaska has accomplished this via the Alaska Permanent Fund. With declining oil revenues, the resolve of Alaskans to keep the principal inviolate may be tested. The board of directors of the fund is already considering erasing the distinction between earnings and principal, which would put the fund at the top of the proverbial slippery slope. In the interests of the greater good, I hope the right long-term decision is made.

Back in the late 1960s and early 1970s, many wanted to keep all the oil in the ground. This would have consigned Alaska to the status quo of a federal- and state-dominated economy, dependent on government payrolls and funding for occasional government building projects. I always wondered why the adversaries thought that status quo was a better option than letting the oil industry explore and produce, a policy that would ultimately pump new money into the economy. The critics seemed to be saying Alaskans themselves would squander the fruits that oil production might bring.

For now, the Alaska Permanent Fund is solid evidence that Alaska has done much to stretch the benefits of Prudhoe far into the future. Unless the Arctic National Wildlife Refuge is explored successfully, North Slope production will end and a serious reassessment of not only the dividend, but also the principal of the fund, will have to be addressed. That will demand calm reason and balanced judgment, not only from the leadership of the state, but also from the electorate. On December 31, 2005 the balance in the fund was 32.2 billion dollars. On June 28, 2007 the balance in the fund was 36.3 billion dollars.

In April 2000, the *Prudhoe Bay State* No.1 was added to the National Register of Historic Places. The response of local leaders representing environmental groups, the Sierra Club, Friends of the Earth, and many others of like mind, was sour grapes. Nonetheless, their members, both then and now, continue to use the conveniences provided by the discovery at Prudhoe Bay. I have often wondered if many of them actually return their annual dividend from the Alaska Permanent Fund, which derives its income from oil. The environmentalists have every right to dislike the decision to honor Prudhoe. I thought the designation was deserved, but I did not consider it a coup. The location was already adequately marked and well known by those in the area. When newcomers visit, they are generally guided there; one cannot imagine a guide ignoring the site, with or without the Historic Places designation.

In the late 1960s and early 1970s, many environmental groups held meetings, usually in opposition to the proposed pipeline. At one such meeting, sponsored by Friends of the Earth, an individual arose and suggested that the group promote National Historic status to the Iditarod Trail. The trail is a major transportation route for dogsled teams and home of the nationally acclaimed dogsled race from Anchorage to Nome, a former gold mining town. The environmentalists waxed eloquent about this proposal for some time. When the discussion had about run its course, I rose and said that I was shocked and stunned that this environmental group would want to accord such status to a trail that was the vestige of so crass an industrial enterprise as gold mining. I opined that as an employee of the oil industry, I personally thought it a good idea, because perhaps 100 years from now, the group might consider proposing national historic status for…(long pause)…the Trans-Alaska Pipeline. The meeting was adjourned.

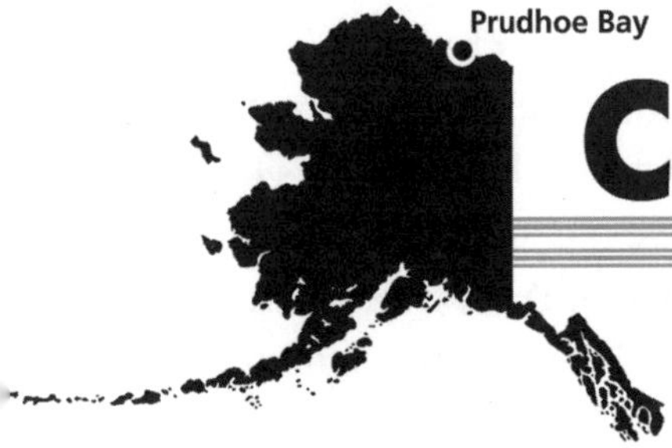

Conclusion

I visited with Harry Jamison in early 2006 and we talked about the spin off of the discovery at Prudhoe Bay. It is difficult to convey its peerless magnitude. It is still amazing that so many different factors converged at this one place during the formation of the earth. All the elements were at or near the apex of good possibilities. Our success at Prudhoe set in motion an avalanche of beneficial changes that were unforeseen then, and still reverberate in Alaska and the nation today.

The discovery became the catalyst that finally brought about the long-festering resolution for Alaskan Natives, who for years had been seeking restitution for the taking of their lands. The Native Claims Act provided many avenues for them to direct their own destiny and to become a major influence in State affairs. It was the need for the pipeline that provided the leverage for their political clout.

In the late 1960s the Santa Barbara oil spill awakened the environmental movement. The pipeline became their prime target and their influence focused regulatory attention on problems of geology, climate and terrain that required new engineering solutions never before tackled on so grand a scale. These problems had to be solved or mitigated to the satisfaction of an enormous variety of national stakeholders.

The sheer size of the discovery necessitated a pipeline of a magnitude that made it one of the greatest construction efforts in history. The actual construction was daunting, but the pipeline has now been in place so long with so few problems it is taken as a matter of course.

The impact of the discovery on Alaska has been profound. The economic impact is almost beyond comprehension. For the oil industry it has been a two-way street with huge streams of money in and out of the operation. What I have mentioned in the text probably only scratches the surface. The State of Alaska Permanent Fund is still growing and dispensing to Alaskans on an annual basis.

All oil fields are finite, and the death of the golden goose will come some day. I wish the leaders in Alaska good fortune in dealing with what is to come. ANWR may be a possible stopgap if it is opened and produces. I believe the oil situation in the U.S. is so critical that sooner or later our congressional leaders and the president will have no choice but to explore ANWR and the Outer Continental Shelves, now closed to entry, where great potential also exists. Even with success, I think in the next quarter century the energy situation will dictate a different lifestyle than we know now.

How fortunate we have been that the national interest in 1967 and the following years was such as to allow for the discovery and production of the Prudhoe Bay field, so that in 2006 it still provides us with 17 percent of our domestic crude. Are there other Prudhoe Bays (or other types of discoveries) in our future? Only time will tell.

Several years ago I heard there were arguments about the actual date of the discovery. I was not involved in that, but many bits and pieces had to fall in place after February 7, 1968 before we knew we *had* a discovery. The testing of the *Sag River State* No.1 in the Sadlerochit on July 17 filled the gap. To me it all came together July 22, 1968 when Robert O. Anderson wrote to the stockholders and the world that ARCO-Humble made the *Discovery at Prudhoe Bay*. The internationally recognized firm of DeGolyer and MacNaughton estimated five to ten billion barrels of recoverable oil.

As I write these final few words, I am thinking collectively of all those still present and those passed on who were involved in Prudhoe Bay. I again dedicate this in honor and in memory of all of them for the parts they played. It took a big team to pull it off. I salute all of you.

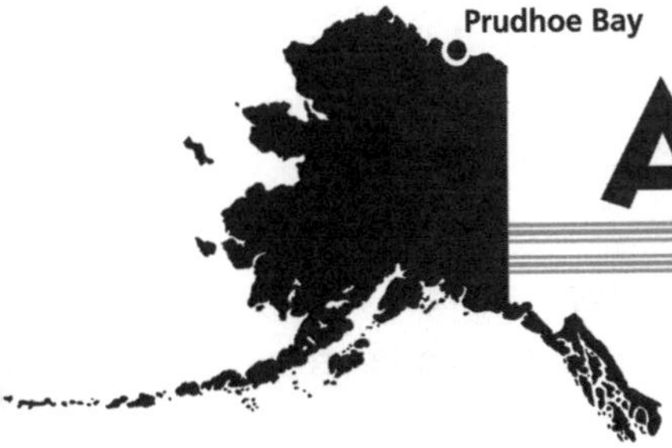

Appendix

Foran Field Journal

(Author's Note: spelling, style and grammar are as they were in original copy. When I have made any comments, they are inserted in square brackets.)

SEPTEMBER Sounds late to be in this neck of the woods with only a summer outfit.

1st – Cold–Cloudy–Snow
Started moving up small valley toward lake, moved 5 miles. No rock exposures–ten ptarmigan–no caribou. Grub situation getting serious. Hope ptarmigan hold out. Camped a few miles below branch to east. Wix took rifle to hunt for caribou, but saw nothing rest of party think we are either doomed to starve or freeze to death. Belgard happy, not much cooking to do. Hughes and Belgard rip sleeping bags and make a double bag to keep warm. Everybody sleeps with clothes on. No one but a brave man would attempt washing his hands or face in the icy water. He would keep shivering the rest of the day. Men can't get minds off food problem. Muting (sic) [He probably meant mutiny.] imminent????

2nd – Snow & Cold
Moved past forks and near lake, could not make Lake. Saw several

exposures and studied structure. Belgard did his stuff again today and bagged 19 ptarmigan. Food getting exceedingly low.

3rd – Cold–snow & cloudy.
Mushed camp & canoe to lake, arrived Lake 8:30 P.M. Tired & weary but more confident Nimiuktuk must head across ridge from lake. Will find out tomorrow. Camped on north side of lake.

4th – Cold, snow & cloudy.
Food supply so low that something had to be done regarding a new supply, or hike back for Icy Cape, and take a chance on catching a late boat out. Sent Belgard and Hughes back to food cache in canyon. Instructed them to mush long hours and sleep with ample clothes using ½ sleeping bag between them. They started back about 9 o'clock: at the same time Wix and I took a small lunch and started southward over passes and hills in search of Noatak drainage. We saw none by 6 P.M.–Wix returned to camp at 10 P.M. tired and hungry. I had kept on moving southward hour after hour, hoping that I might find a stream flowing southward–finally, about dusk, I came on a lofty peak on the crest of a high group of mountains. Looking south and down a steep walled valley I saw the head waters of the south branch of the Colville that I crossed about 4 P.M., and the Nimiuktuk that we had been looking for. I went down into the valley to make sure. At 10 P.M. hungry and tired, having had nothing to eat but a pilot biscuit and 3 sardines at noon I started back down the winding Colville Fork to the point at which I crossed earlier in the day. It was freezing and a heavy frost in the air formed clouds. At 3 AM I passed the crossing place, the frost-fog masking any landmarks by which I might recognize the route I traversed in the afternoon. I walked until about 4 AM–passed several unfamiliar spots; including a couple fair sized streams from the south, that I was sure must have been to the left of me on the journey in the afternoon. I turned, and retraced my steps for two

hours, then crossed the river, went over the ridge and at 8 AM arrived in the valley occupied by the lake, but about 3 miles east of camp. Made this three miles in about an hour and a half. When I arrived, frost in my beard and hair, tired and hungry. Lonseth and Wix had packs with supplies on their backs ready to search for me. They were as glad to see me as I was to see them. Home could never compare with the tent and sleeping bag at this genesis into camp. Feet swelled, and very sore. Had new pair of army doughboy shoes on, rubber boots worn out from wear on sharp rocks in rivers. Had cup cocoa, corn and climbed into sleeping bag.

10 P.M.

Sept. 5th – Warm and clear.
Awake, Wix and Lonseth had supper cooked. They had just gotten in from packing a load each, to the other branch of the Colville to the south.
Ate supper 10:30 P.M. doctored feet and went to bed again.

6th – Cool–cloudy of thick heavy frost on tundra. Tied ropes to canoe and wind gust & I dragged it 2/3 of way to first divide. Feet still swelled but not so sore. Heel and ankle skinned, very unpleasant. Looking for boys with new grub supply. Did not show up.

7th
Lonseth, Wix and I packed part of camp to divide. On returning to camp found that boys had returned with 60 lbs. of food. Both of them tired and asleep in tent when we arrived. They had a miserable, hard time, no sleep and bad weather. Whole party to bed early to be ready for uphill pack in morning.

8th
Broke camp on lake 7 A.M. Each man carried huge load to Colville

branch. Wix stayed and fixed supper; rest of us went back and got canoe. Very cold in evening, no wood, and stream frozen over. Men all rustled small stems and particles of wood–enough to cook supper and breakfast. Absolutely, no further wood with 20 miles–all wood used at lake camp was brot from main branch of Colville. Country covered with one inch of frost. Weather so cold and fire wood so scarce, that water in kettle would freeze before it could get warm enough to combat temperature. Fire would die down before heat was affective. May have to use tent poles and personal effects to make fire for cooking. Men all use sleeping bags to keep warm when not working. Nobody dares take off his clothes for fear of freezing to death before he can climb into his sleeping bag. It is quite evident that we are not in a climate suited to summer clothing.

9th – Very cold and cloudy
Only a handful of twigs to cook three days food. Used two tent poles, tripod legs, canoe paddle and "Elto" oil to cook enuf food on which to go three days. Each man packed one load across mountains thru low pass to south drainage.

Bed of conglomerate exposed at crest of ridge, exposed again ¼ mile south of low pass. A great overthrust occurs at crest of ridge between lake and camp. New formation of dark colored sediment exposed from Lake Valley to chert of over thrust. Explanation of the geology in front of book.

Mts. very irregular and pretty, climate much warmer on south side of mountains. Like dropping out of northern Alaska into California.

10th – Cold–Frosty
Plowed [packed?] another portion of camp across pass. Cached transit–alidade undershirt–towels–all leather shoes–large binoculars–15 pounds cooking utensils–rock hammers–cache cover–fossils–rock samples–note books–celluloid etc.

11th – Cold & Frosty
Broke camp and moved important part of camp 15 miles down Nimiuktuk to good wood supply–large cotton wood and willows. Had large fire and celebrated arrival on Noatak water shed. River still freezes during night, water getting scarce in Nimiuktuk as river is descended.

12th - Cold–Snow & Cloudy
Packed part of camp left at pass to camp, packed canoe half way to camp. Wix took load 6 miles south of camp to see how close we were to Noatak. Can't see Noatak.

13th
Broke camp–moved about 6 miles down river to large fork to west. Water ordinarily would increase in quantity but the weather in the mountains getting colder day by day diminishes the flow of water in the streams. Too late in season for canoe transportation on subordinate streams. Food nearly played out again. No game to be had. Camped at forks.

14 – Cold & cloudy.
Moved camp 6 miles. Can work only 4 hours on small food rations–last pot of beans. Tomorrow will end peas & rice–have had no coffee–butter–sugar–milk for week. Morale of party very low. Belgard (cook) would have vacation except for back packing.

15th – Cold & cloudy
Had peas & rice for breakfast, steero [He uses cube with it at the end. Mo gives me broth from a cube heated in water sometimes.] soup for lunch–peas and rice for supper–no more food except steero and bacon rinds.

Moved camp 4 miles – Belgard, Hughes and I each packed large load, Wix and Lonseth put rest of camp in canoe and moved on water.

16th - Cold & Rainy [A 1924 happy face penciled in notes]
Bacon rind is, and will be the menu until our arrival at Mission on Noatak. Broke camp below main forks of Nimiuktuk put entire camp in canoe and moved to Noatak. Men so thrilled that absence of food is only secondary matter. Stomachs in first class shape but we cannot work over 3 hours per day, no stamina. Camped at old cabin at mouth of Nimiuktuk.

17th – Cold & Rainy
Had an early start, made old trapper cabin at spruce timber at upper end of canyon. Could have gone thru canyon but rocks and swift water did not look inviting to men of our condition. Too weak to move quickly on emergency. A wandering seagull flew too close to the boat. I took about 5 minutes aim at it and when it got within ten feet of the boat I shot it. This was a life saver–tasted like chicken. Boys had words with Belgard as to the mysterious disappearance of the seagulls gizzard (every morsel counted heavily at this time). Belgard explained that gizzard was put on Wix plate, but Wix accused Belgard of wolfing gizzard–while cooking the gull. Wix giving Belgard extremely vitriolic purge. Words about hot lead to assist digestion of gizzard. Argument more serious than Dan Magrew's poker controversy. Hughes ordered to sit on gun and ammunition. Men not speaking.

18th – Very Cold & Cloudy–Windy
Up to an early start feeling O.K. Low water in river made canyon rather treacherous. Hughes at helm, proved himself, many times, to be first class river navigator. Risky business, 5 men with summer camping equipment in one small canoe, shooting rapids, some large, bad boulders exposed, and some, just a few inches under water. Wix took the bow and gave Hughes directions–Wix played the role of first mate in commendable fashion. At 11:30 A.M. we stopped and inspected cabin #2 about and mile above Kelly River. Effects of big

meal wearing off, men exceedingly hungry. We know we'll soon arrive at an Eskimo mission, possible tomorrow morning, barring accidents & river not so swift, but shallow in places. Canoe gets broadside with current now and then; this is a dangerous situation, nearly turned turtle several time. Lonseth jumped into deep water once to save boat. Canoe so heavily laden that it draws water nearly to gunwales only four inches to spare. Canoe into Eskimo camp just as we were preparing to stop for steero. A grand and glorious feeling. First we ate raw caribou, then cooked caribou–biscuits–sugar and coffee tasted wonderful to us. Were without it for two weeks. Filled once more and bound for mission. River wide, and wind making it very rough. Canoe leaking badly, and taking waves overboard. Had to stop and camp. Saw whole family of red fox on river bank. Had caribou and tea for supper. 70 miles today. Wix and Belgard renew seagull gizzard controversy.

19th – Very cold & cloudy.
Breakfast, caribou & tea, broke camp at 9 AM–wind died down. Came to fish camp, took about 30 pounds of trout and grayling. Arrived at mission 11 A.M.– ate steadily for 11 hours. Went to bed feeling groggy from eating too rapidly. Withal food Wix still nasty to cook about eating gizzard.

20th – Cool & Cloudy
Nobody slept during night, stomachs out of order–too much sweet food. Every man in part has bad case of diarrhea. Took new supply of food, had bottom of canoe repaired, left Mission 11:30 A.M. Had to stop canoe so many times, due to mens bowel disorder that we made only 25 miles –camped overnight 3 miles below canyon. Took all sizes & shapes of pills–cannot get stomachs in regular order. Men to hungry to ease up on eating. Hughes stomach is very bad condition, will have to go to gov't hospital in Kotzebue or Deering. Belgard accuses Lonseth of eating gizzard.

21st – Cold, rainy & windy
Hughes stomach in pretty bad shape. Lucky we are only short distance from Kotzebue. We have no medicine that will aid him. Looks thin and pale, lower limbs puffed out like attack of dropsy. Broke camp, arranged bed in canoe for Hughes, resting comfortably. Started en route for Kotzebue 8 A.M. were exceedingly lucky in getting across Inlet before severe storm. Arrived Kotzebue 11:30 A.M. had nurse, Miss O'Brian attend Hughes, says his blood is polluted with sugar and stomach is overworked. Says that light meals and rest will improve his health.

All hands took dinner at Molly's R.H. Told the camp of our trip. They were surprised to see a canoe with five white men come down the Noatak so late in season. Traded excess grub for quarters in cabin. Received a telegram from Mertie, answered it, asking him to make some arrangement for our transportation to Nome, also wired to Dr. Brooks, received answer from G. O. Smith. Hope nothing has happened to Dr. Brooks, health bad when I left Washington in spring. He and Paige last men I saw in D.C. before leaving. Looks as tho we will have to mush to home from Deering. Trying to charter boat at Kotzebue. All navigation closed on Arctic Ocean. Lemen's refuse to send "Silver Wave."

Sept 22 to Sept 30
Lonseth, Wix and Belgard in furious row trying to solve mysterious to disappearance of gizzard. Belgard gave rational statement of his integrity. Could have used steero cube but didn't. Had a good rest in Kotzebue. Hughes feeling well again but cannot get swelling out of lower limbs. Trader Paul Davidekitz cleared gizzard altercation–Seagull is scavenger, has stomach–no gizzard. Wix & Belgard reconciled. couseth??

After sending about 8 telegrams we finally got the "Kobuk" 12 ton gas boat owned by Sam Magids to take us to Nome. Wix & Belgard friends again.

Oct 1st to 3rd

Boarded "Kobuk" 11 AM Oct 1, bound for Nome–stopped at new fox farm at Shesmeref 8 AM. 2nd pulled out 9 AM–hit storm at Cape Prince of Wales and fireworks began. Engine trouble caused us to drift back into Arctic again. No room to sleep in cabin, Lonseth and I tie ourselves to top of boat, but both of us retire to cabin soaking wet–sleeping bags also saturated. No sleep–but we will soon be in hotel. Storm subsided in morning of 3rd, engine working again arrived in Snake river harbor 4:30 PM Oct. 3rd. Three Nome detectives and marshal aboard boat in harbor and search for liquor. They completely ransacked the baggage, found nothing. Glad to be in Nome again. Looked up Mertie and wired to Washington.

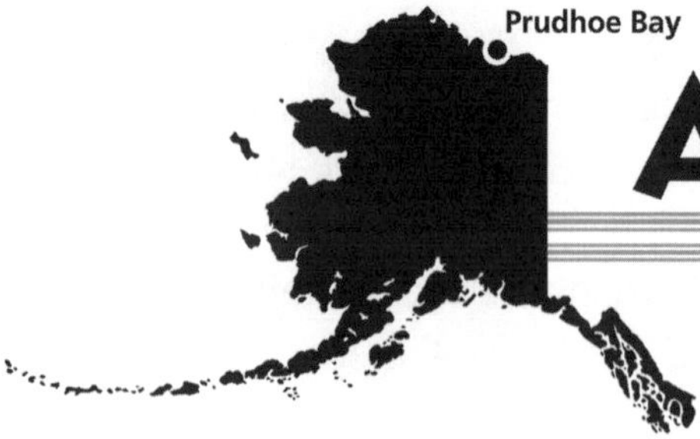

Appendix

Honorees

Most of the people in the following list of names were in the continuum of activities, which began in 1899 and culminated in the drilling of the discovery and confirmation wells. It is to them this history of the discovery is dedicated. Those who were not in the chain of events leading to the discovery were associated with it in some way in the later stages of the discovery. When one makes a list of names there is a great risk of overlooking someone. I hope I have not done that.

R.D. Adams
Robert S. Agatston
William Albright
William Allen
Samuel Allen
Ann Byrd Allen
Bob Anderson, geologist
R.O. Anderson
(Chairman) Joe Arndt
G. Ray Arnett,
Ernest H. Arp
J.R. Arrington
Tatiama Ashhurkoff
A.G. Austin
Julius Babisak
Forest Baker
J.R. Balsley
W.G. Banks
L.G. Bardin
Robert E. Barker
W.L. Barksdale
W.M. Barnwell
Thomas Barrow
Kerby Bay
Harold Beaver
R.W. Belgard
Roscoe E. Bell
Henry Bender
Jack B. Bennett
Carl Benson
E.M. "Mo" Benson
Harlin R. Berdquist
Wenonah E. Berdquist
Rudy Berlin
W.H. Berry
Douglas Bertek
R.S. Bickel
William Bicknel
W.S. Bicknell, Jr.
William C. Bishop
Walter R. Blankenship
Charles Boles
Paul D. Bollheimer
Chet Bonar
Arthur L. Bowsher Sr.
Carl Bradey
Thornton F. Bradshaw
Thomas Brady
Lonnie L. Brantley
Arnold Breckenridge
Lester D. Brockett
O.D. Brooks
Richard Brooks
W.P. Brosgé
Roland W. Brown
Leo Brown
Allen Brown
Roland Brown
Bert Brown
Lt. (j.g.) S.C. Brown
Donald Bruce
Creighton Burk
R.J. Burnside
Ernst H. Bush

Henry A Campbell
William Campbell
Jack Carlisle
Robert Carlsen
John Carr
Matthew V. Carson
W. Douglas Carter
John Cavanaugh
H.B. Chambers
Roland Champion
Robert M. Chapman
Peter Clara
Jack Clark

George Clark
Fred Clark
James Clinton
Robert R. Coats
W.A. Cobban
J.B. Coffaman
Bill Congdon
Jack Cook
Lt. (j.g.) H.C. Cortes
Ralph F. Cox
Louis Cramer
Richard W. Crick
W.H. Crocker
A.A. Curtin
R.C. Curtis

Max H. Davis
Louis F. Davis
Gordon Davis
Morgan Davis
Frank Deinzer
T.F. Derrington
James De Santis
R.L. Detterman
Ome Diaber
Angela Ditch
Frank B. Dodge
William L. D'Oiler Jr.
G.C. Donahue
J.H. Downs
Richard G. Dulaney
Helen Duncan
D.H. Dunkle
Joseph Dunn
J.T. Dutro
S.E. Dwornik

Norman Ebbley
Carl H. Ebbley
C.D. Eberlein
Rolin Eckis
William Eiting
G.W. Engle

Charles Fabian
Leo Fay
Allen Feder
R.E. Fellows
Walter Fillipione
Frank K. Fish
Lt. Gerald FitzGerald
Stuart Folk
William T. Foran
Sir John Franklin
N.B. Frickman

O.K. Fuller

Bert Gailbraith
Joe Galant
J.H. Galloway
W.R. Gardner
Philip Gaston
George O. Gates
Peter Gathings
Roger Gearing
L.T. Gentilomo
C.E. Giffin
King Gilette
James Gilluly
MacKenzie Gordon
Don Grinsfelder
Zed Grissom
George Gryc
E.C. Guerin

Merrill W. Haas
Robert J. Hackman
H.W. Haight
Rodger D. Hamilton
W. Dow Hamm
Dale A. Hauck
J. Heemann
W.E. Henrichs
Donald Henrickson
Les Herndon
Gordon W. Herred
Walter J. Hickel
G.B. Hickok
Mason Hill
George Hipple
John Hipple
Dr. David Hite
Phil Holdsworth
Judy Hopkins
Ray Horton
W.S. Howard
Lt. W.C. Howard
H.G. Hughes
Mr. Hunt
T.S. Hutchinson

Ralph W. Imlay
Jack Irwin
Delmer L. Isaacson

J.R. Jackson
D.L. James
H.C. "Harry" Jamison
Red Jenkins

Linda Jennings
Homer Jensen
Donald Jessup
N.J. Johnson
Norma Johnson
Robert Johnstone
D.L. Jones
C.H. Jones
Crandall Jones
N.B. Jones
Don Joplin
Henry Josling

James Keasler
James Keeler
William Keeler
A.S. Keller
Landon Kelly
B.H. Kemmel
B.H. Kent
William Kieschnick
Jim Kiesler
S.W. Kingsbury
C.E. Kischner
Joe L. Klutts
Clarence Knutson
A.N. Kover
Rodney Kraus
Lt. W.L. Kreidler
Paul D. Krynine
Irv Kunz
Jack Kurtz

A.H. Lachenbruch
M.C. Lachenbruch
B.J. Lancaster
Geoff Larminie
E.H. Lathram
Bennie Laudermilk
Nell Lawrence
Elder Lebert
E. de K. Leffingwell
J.A. Legge
L.B. Lehrs
John Levorsen
Jorgen Lilliebjerg
Helen Tappan Loeblich
Audrey Loflus
John Loftis
H. Lonseth
Robert Lord
Frank Love
J. Stewart Lowther
Howard Lucas

L.B. Luhrs
Richard K. Lynt
Paul Lyon

Marvin D. Mangus
Joe Mann
I.W. Marine
Phil Marsh
Robert J. Marshall
Thomas Marshall
Lt. Ernst Marti
D.E. Matthewson
Jack McCarney
Lt. (j.g.) A.P. McConnell
R.C. McGregor
Samuel McIntire
F. Stearns McNeil
Frank E. McPhillips
Leland Megason
John Calvin Merrit
A.V. "Tony" Messinio
J.B. Mertie Jr.
C.R. Metzer
W.H. Meters
Andrew Milek
James Minick
Jim Moore
William B. Moore
Richard Moore
Joe L. Morgan
Dean Morgridge
Robert H. Morris
Charles T. Morrow
Charles "Gil" Mull
E.J. Mumms
John Murdock
Stuart Mutt

Milton Norton

Richard D. Olson
Alex Osanik
C.R. Overly

Sidney Page
Lt. M.V. Paine
Arnold Palenske
Bert Panigeo
Bryce Parker
Gar Passel
Robert M. Parks
H.G. Patrick
Thomas Patrick

C.R. Patterson
Edward L Patton
W.W. Patton Jr.
Thomas G. Payne
William C. Penttila
W.J. Peters
D.E. Peterson
Elizabeth Peterson
"Pete" Peterson
W.H. Phillipps
Larry Pipes
Hary Pistole
Calvin Post
H.B. Post
Wallace Pratt
M.R. "Proc" Proctor

R.G. Ray
D.E. Reed
John C. Reed
J.B. Reeside Jr.
H.N. Reiser
Carl Reistle
Henry Repp
C.D. Reynolds
Jim Reynolds
H. Glen Richards
Gene R. Richards
T.L. Richardson
Leslie Riggins
Robert Ritter
Thomas G. Roberts
Buford Robinson
Florence Robinson-Weber
Lt. (j.g.) J.A. Rogers
J. Rosebaugh
William Rosser
John Rossieter
John Rousi
Norman J. Ruback
Gen Ruby
Florence Rucker-Collins
Henry Rupp
C. Russell
Ben Ryan
Irene Ryan

Edwards G. Sable
A.B. Sanders
Verna Sandison
James Savage
William Scheidecker
Harold Schetter
Bill Schetter
Chuck Schierck
S.P. Schoonover
Frank Charles Schrader
Fred Schultz
Walter Scott
G.L. Scott
Charles Seagle
Albert J. Seeboth
Charles H. Selman
Mickey Sexton
Vincent Shainin,
Paul H. Shannon
J. Norman Shear
Bill Shetter
I.G. Shon
Ken Siebel
Charles Siegel
John Sindorf
O.L. Smith
Harold Smith
A. Malcolm "Sandy" Smith
P.S. Smith
Walter R. Smith
John Smith
W.B. Smith Jr.
I.G. Sohn
Ralph Solecki
Fred Sollars
L.A. Spetzman
Armand Spielman
P.W. Stade, Jr.
R.K. Steer
Karl Stefansson
John M. Stevens
Lamar Stone
James Stover
Betty C. Strah
Henderson Supplee
K.L. Suydam
F.M. Swain

Robert Tabbert
I.L. Tailleur
Harry A. Tait
E.F. Taylor
Vic Tessier
A.N. Tetraulat
Gerald "Jake" Thomas
John Thomas
Henry Thompson
Capt. J.S. Thompson
R. M. Thompson
Robert F. Thurrell
Henry Time
"Mandy" Touring
J.T. Townley III
Rex Townsend
D.M. Trainer
Bill Travers
M.L. Troyer
Thomas Tucker
W.W. Turnbull
J.C. Turner

Clarence Unruh
M.E. Van Cott
William B. Van Valen

Ernst H. Wadsworth
Lt. (j.g.) G.S. Wagman Jr.
Robert K. Walker
Wray Walker
Bill Walters
Loren Ware
L.A. Warner
James Watkins
John Watson
Michael Webb
E.J. Webber
C.C. Wemberly
Frank Whaley
James Whitaker
D.A. White
Henry A. Whitley
C.L. Whittington
F. Eugene Wiancko
John Wiedman
William Wiggan
Neal D. Williams
Lee Wilson
James L. Wilson
G.R. Winter
O.L. Wix
E.F. Wolf
J.P.W. Woods
L.D. Woody
M.A. Wright
Les Wunsch
R.E. Wycoff

Ellis Yochelson

J. H. Zumberge

USGS Personnel During NPR4 Exploration

Samuel Allen
Titiama Ashurkoff
A.G. Austin

J.R. Balsley
W.G. Banks
L.G. Barbin
W.L. Barksdale
Henry Bender
Carl Benson
Harlin R. Bergquist
Wenonah E. Bergquist
Douglas Bertek
R.S. Bickel
A.L. Bowsher
W.P. Brosge
Roland W. Brown
Leo Brown
W.L. Brown, Lt.

Matthew V. Carson
W. Douglas Carter
H.B. Chambers
Robert M. Chapman
Robert R. Coats
W.A. Cobban
Florence Rucker Collins
Jack Cook
H.C. Cortes, Jr. Lt.
A.A. Curtis

Ome Daiber
Frank Dainzer
Max H. Davis
T.F. Derrington
R.L. Detterman
O. Diaber
Wm. L. D'Oiler, Jr.
G.C. Donahue
J.H. Downs
Harald D. Drewes
Helen Duncan
D.H. Dunkle
J.T. Dutro
S.E. Dwornik

C.D. Eberline
G.W. Engle

Allen Feder
R.E. Fellows
Walter Fillipione
W.A. Fischer
Frank K. Fish, Lt. USN
Sturart H. Folk
W.T. Foran, Lt. USN

George O. Gates
L.E. Gentilomo
MacKenzie Gordon
George Gryc

Robert J. Hackman
Dale A. Hauck
Marvin Heany
W.E. Heinrichs
Gordon W. Herred
J. Hipple
G. Hipple
W.S. Howard

Ralph W. Imlay
Jack Irwin
Delmer L. Isaacson

"Red" Jenkins
Homer Jensen
D.L. Jones
Don Jopling
Henry Josling

A.S. Keller
B.H. Kemmel
B.H. Kent
C.E. Kirschner
Clarence Knutson
A.N. Kover
W.L. Kreidler, Lt.
Paul D. Krynine
Jack Kunz

A.H. Lachenbruch
M.C. Lachenbruch
E.H. Lathram
Norman Lbbley
Elder Lebert
J.A. Legge
L.B. Lehrs
Helen Tappan Loeblich
Audrey Loftus
J. Stewart Lowther
Howard Lucas
L.B. Luhrs

Marvin D. Mangus
I.W. Marine
Ernest Marti, Lt. USN
D.E. Matherws
D.E. Mathewson
Jack McCarney
A.P. McConnell, Lt. Jg
R.C. McGregor
F. Stearns McNeil
Robert H. Meade
C.R. Metzger
W.H. Meyers
Andrew Milek
Jim Moore
Robert H. Morris
Charles T. Morrow
E.J. Mumms

William L. Nystrom

Richard. D. Olson
Richard D. Olson

M.V. Paine, Lt.
Arnold Palenske
W.W. Patton, Jr.
Thomas G. Payne
D.E. Peterson
W.H. Phillippi, Lt.
Calvin Post
H.B. Post

R.G. Ray
D.E. Reed
J.B. Reeside, Jr.
C.D. Reynolds
H. Glen Richards
H.N. Riser
Thomas G. Roberts
L.A. Rogers, Lt. Jg
J. Rosebaugh
Glen Ruby

Edward G. Sable
A.B. Sanders
Charles Schierck
S.P. Schoonover
Charles Seagle
Vincent Shainin
Paul H. Shannon
Charles Siebel
Harold Smith
O.L. Smith
I.G. Sohn
Ralph Soleck
Lloyd Spelzman
Karl Stefansson
John M. Stevens
K.L. Suydam
F.M. Swain

I.L. Tailleur
E.F. Taylor
Vic Tessier
A.N. Tetraulat
J.S. Thompson
R.M. Thompson
Henry Thompson, Capt. Robert F. Thurrell
J.T. Townley III
Rex Townsend
M.L. Troyer

Ernest H. Wadsworth
G.S. Wagman, Jr. Lt. Jg
Jack Warne
L.A. Warner
John Watson
E.J. Webber
Florence Robinson Weber
C.C. Wemberly
D.A. White
C.L. Whittington
F. Eugene Wiancko
William Wiggan
G.R. Winter
E.F. Wolf
Ellis Yochelson

J.H. Zumberge

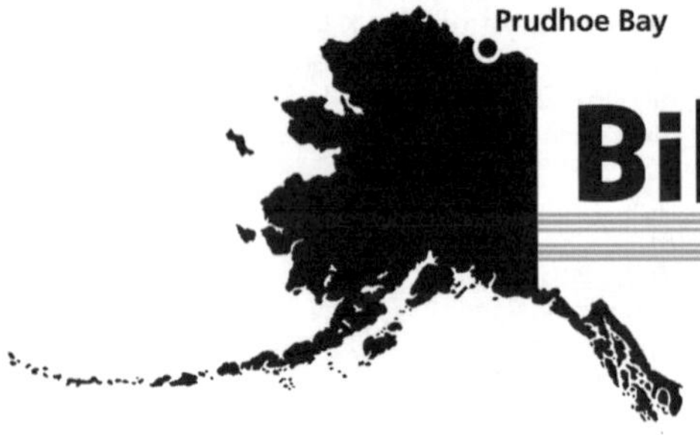

Bibliography

Alaska, State of Permanent Fund. *Fiscal Year First Quarter Report.* 2003: 14.

Alaska, State of Permanent Fund. *Dividend Income 1982-2005.*

Alaska, State of. Department of Natural Resources, Division of Oil and Gas. Exploration wells for Alaska. 1988: 9.

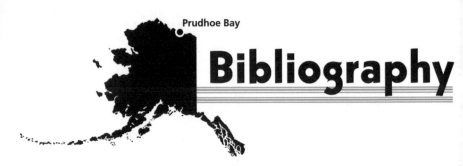

Alaska, State of. 1968. Department of Natural Resources, Division of Oil and Gas. Prudhoe Bay State Number 1 Well file, numerous reports, letters and documents.

Alaska Oil and Gas Commission. 2000. *Alaska Production Summary by Pools*: December 2000, November 2001, October 2002: 7.

Alaska Weekly. 1923. Discoverer Smith Tells of Finding Petroleum on the Border of Arctic Ocean. April 20, 1923. (Part of Appendix Pre–PET 4—Adams Expedition Cape Simpson, 1921).

American Geological Institute. 1976. *The Dictionary of Geological Terms*. Falls Church, VA: Anchor Book.

Anderson, R.O. and Bradshaw. 1968. Letter to shareholders. July 22.

ARC Energy Charts. 2006. January 9: 11.

Arnett, G. Ray. 1957. Business letter to William C. Bishop, Richfield Oil Corp.

Arnett, G. Ray. 2000. Email to John Sweet. August 13.

Babisak, Julius. An Atlantic Richfield History Project. Unpublished personal communication. 1979: 6.

Babisak, Julius. 2000–2001. Various personal communications.

Badger, T.A. An ARCO Epitaph. *Alaska Magazine*. August. 2000: 30–33, 67.

Baker, Frank E. 2005. Improved oil recovery method to boost Alaska production. *BP Horizon*, April: 48.

Bates, Robert L. and Julia A. Jackson. 1990. "Glossary of Geology." American Geological Institute.

Bazar, John. 2000. *Yukon Alone*, New York: Henry Holt.

Blocher, R.L. 1957. Developments in Alaska 1956. *American Association of Petroleum Geologists Bulletin*, vol. 41, 6 June: 1311–1315.

Brosge, William P. and Charles L.Whittington. 1966. Geology of the Umiat–Maybe Creek Region, Alaska: *U.S. Geological Survey, Prof. Paper*, 303–H: 501–632.

Bowsher, Arthur L., and J.T. Dutro Jr. 1957. The Paleozoic section in the Shainin Lake area, Central Brooks Range, Alaska: *U.S. Geol. Survey Prof. Paper* 303–A, pt. 3: 1–39.

Bowsher, Arthur L., 1987. Sinclair Oil, and Gas Company on the Alaskan North Slope, 1957–69. *Alaskan North Slope Geology,* Vol. One. The Pacific Section, Society of Economic Paleontologists and Mineralogists, and The Alaska Geological Society: 13–18.

British Petroleum, *Horizon.* 2006. Pipeline milestone. February: 8.

British Petroleum *Horizon.* 2007. Gas hydrates. A viable frozen resource? By Frank Baker. April: 26

Brown, Allen. 2000. Personal communication.

Chapman, Robert. Unpublished. List of people who worked on the NPR4 USGS field 69. *Alaskan North Slope Geology,* Vol. One The Pacific Section, Society of Economic Paleontologists and Mineralogists, and The Alaska Geological Society. 1989: 13–18.
Chapman, Robert M., and Edward G. Sable. Geology of the Utukok–Cortwin Region, Alaska. *U.S. Geological Survey Prof. Paper 303–C.* 1960: 47–162.
Chapman, Robert. M., R.L. Detterman and M.D. Mangus. Geology of the Killik–Etivluk Rivers Region, Alaska. *U.S. Geol. Survey Prof. Paper 303–F.* 1964: 325–407.
Cooper, Bryan. 1972. *Alaska — The Last Frontier*, Hutchinson & Co. LTD.
Coppinger, Raymond and Lorna. 2001. *Dogs — A Startling New Understanding of Canine Origin, Behavior and Evolution*, New York: Scribner.
Crick, Richard W. 2000. Email with the author. 9 April.

De Golyer, E. Wallace. "Everette Pratt First Sidney Powers Memorial Medalist: An Appreciation," *American Association of Petroleum Geologists Bulletin* vol. 29 Number 5. 1945: 477–487.
Detterman, R.L., Robert S. Bickel and George Gryc, Jr. "Geology of the Chandler River Region, Alaska." *U.S. Geol. Survey Prof. Paper 303E.* 1963: 223–324.
Dutro, Jr. J. Thomas. "A Brief History of Eighty Years of Geological Exploration in the Central Brooks Range, Northern Alaska," *Alaskan North Slope Geology,* Vol. One The Pacific Section, Society of Economic Paleontologists and Mineralogists, and The Alaska Geological Society. 1987: 7–12.

East, Edwin H. "Developments in Alaska in 1966." *American Association of Petroleum Geologists Bulletin*, vol. 51, Issue 6. 1137–1151.
"Energy & Resources: American Geological Institute Energy Notes." *Geotimes*, vol. 46, no. 11. June 2000.
Ewing, Susan. 1996. *Alaska Nature Factbook*: Alaska Northwest Books.
Exxon. "Growth in a Changing Environment," *A History of Standard Oil Company (New Jersey) 1950–72 and Exxon Corp. 1972–75.* 1975: 134–145.
Exxon House Magazine. Search vol. 2 no. 1 Sept. 1986: 6–13.

Fay, Leo F. "Developments in Alaska in 1965." *American Association of Petroleum Geologists Bulletin*, vol. 50, Issue 6. 1966: 1311–1323.
Fleming, Fergus. 1998. *Barrow's Boys*. New York: Grove Press.
Foran, William T. 1924. Field Journal from 1924 field party. Appendix Early Pet–4 USGS.
Francis, Devon. 1985. *The Power to Fly.* USA: Aero Publishers.
Franklin, John. 1828. Narrative of a second expedition to the shores of the Polar Seas in the years 1835, 1826 and 1827. London.
Fritz, Mary. "A Giant Who 'Went Blithely.'" *American Association of Petroleum Geologists, Special Issue* "A Century" 2000:38–39.

Gryc, George. "History of Petroleum in Northern Alaska: Proceedings of the Geological Seminar on the North Slope of Alaska: Pacific Section." *American Association of Petroleum Geologists*. C1–C10. 1970.
Gryc, George. 2002. Personal Email on a 1950 Helicopter Crash, NPR4 Alaska.

Halbouty, Michael T. "Wallace E. Pratt: Human Needs Award." *American Association of Petroleum Geologists Bulletin*, vol. 56 No. 9, 1972: 1885–87.
Hall, Wm. J., D.J. Nyman, E.R. Johnson, J.D. Norton. "Performance of the Trans-Alaska Pipeline in the November 3, 2002 Denali Fault Earthquake." Proceedings of the Sixth U.S. Conference and Workshop on Lifeline Earthquake Engineering, Long Beach, CA, August 2003.
Hanna, G. Dallas. "Oil Seepages on the Arctic Coastal Plain, Alaska." *Occasional Paper of California Academy of Sciences*, no. 38. 1963: 18.

Herndon, Booton. 1971. *The Great Land*. New York, NY: ICS Press.
Hickel, Walter J. 2001. *The Alaska Solution*. Oakland, CA: ICS Press.
Hite, PhD, David. 2001. Personal communication on the evolution of Cone Deltas.

International Petroleum Encyclopedia. Penn Well Publishing Company.

Jackson, J.R. Various archival reports and letters.
Jamison, H.C. 1979. *An Atlantic Richfield History Project*. Tape recorded interview with John Nation and James P. Roscow. Unpublished.
Jamison, H.C. 2001. Personal communication.
Jamison, H.C., L.D. Brockett and R.A. McIntosh. "Prudhoe Bay–a ten-year perspective, Giant oil fields of the decade, 1968–78." *American Association of Petroleum Geologists Memoir 30*. 1980:289–314.

Kaniut, Larry. 1994. *Cheating Death*. Fairbanks, Alaska: Epicenter Press.
Keller, A.S., Robert H. Morris and R.L. Detterman. *Geology of the Shaviovik and Sagavanirktok Rivers Region, Alaska, U.S. Geological Survey Prof. Pape. 303–D* 1961: 169–222.
Kirschner, C.E. "Developments in Alaska 1957," *American Association of Petroleum Geologists Bulletin*, vol. 42, Issue 6, June 1958: 1434–1444.

Leffingwell, E. deK. *The Canning River, Northern Alaska, U.S. Geological Professional Paper 109*. 1919: 251.
Leffingwell, E. deK. "My Polar Explorations 1901–14." *Explorers Journal*, October 1961: 13.
Lian, Harold M. 1960. "Developments in Alaska 1959." *American Association of Petroleum Geologists Bulletin*, vol. 44. Issue 6, June 1960: 940–945.

Mangus, Marvin Dale. An Atlantic Richfield History Project, tape recorded interview with John Nation of ARCO and James P. Roscow, former Business Week reporter: Unpublished. 1979: 47.
Mann, H. 1959. "Developments in Alaska 1958." *American Association of Petroleum Geologists Bulletin*, vol. 43, Issue 6, June: 1427–1436.
Marshall, Robert. 1970. Alaska Wilderness. Berkley, CA: University of California Press.
Masterson, W. Dallam and Chester E. Paris. "Depositional History and Reservoir Description of the Kuparuk River Formation, North Slope, Alaska." *Alaskan North Slope Geology*, Vol. One. The Pacific Section, Society of Economic Paleontologists and Mineralogists, and The Alaska Geological Society. 1987: 95–107.
Mertie, Evelyn. 1982. *Thirty Summers and a Winter*. Mineral Industry Research Laboratory, Fairbanks, Alaska: University of Alaska.
Metzger, Charles R. 1983. *The Silent River*. Los Angeles, CA: Omega Books.
Mikkelsen, Ejnar. 2005. *Mirage in the Arctic*. The Lyons Press.
Miller, Debbie S. 2000. *Midnight Wilderness*. Anchorage, AK: Alaska Northwest Books.
Mull, Charles "Gil." Personal communications.
Murie, Margaret E. 1978. *Two in the Far North*. Anchorage, AK: Alaska Northwest Books.

Oil and Gas Journal. 1988. April 18.
Oil and Gas Journal. 2004. July 6: 43.
Orth, Donald J. *Dictionary of Alaska Place Names, U.S. Geological Survey Prof. Paper 566–567*.

Patton, William W. Jr. and Irvin L. *Tailleur. Geology of the Killik–Itkillik region, Alaska, U.S. Geological Survey Prof. Paper 303–G*. 1964: 409.497.
Penttila, William C. 2000. Personal communication.
Petzet, Alan. 2000. "Alaska operators start Alpine field, take more leases." *Oil and Gas Journal*, vol. 49, Issue 49. October 4.

Phillips, James W. "Name Origins: Prudhoe Bay and Duke Island." *The Alaska Journal*. 1974: 149–152.
Pratt, Wallace E. "Toward a Philosophy of Oil-Finding," *American Association of Petroleum Geologists Bulletin*, vol. 36, No 12. 1952: 2231–2236.

Reed, John C. *Exploration of Naval Petroleum Reserve No. 4 and Adjacent Areas Northern Alaska, 1944–1953, U.S. Geological Survey Professional Paper 301*. 1958: 192.
Roderick, Jack. 1997. *Crude Dreams*. Epicenter Press.
Roscow, James P. 1977. *800 Miles to Valdez*. Englewood Cliffs, NJ: Prentiss–Hall Inc.
Rossbacher, Lisa A. 2002. "Think like a Geologist," *Geotimes*, Geologic Column, vol. 49, no. 1, August.
Rozell, Ned. 2001. *Walking My Dog, Jane*. Pittsburgh, PA: Duquesne University Press.
Rutledge, Gene. 1987. *Prudhoe Bay...Discovery.* Anchorage, AK: Wolfe Business Services.
Rutledge, Gene. 1998. *Prudhoe Bay...Discovery to Recovery!* Anchorage, AK: Wolfe Business Services.

Salvador, Amos. 1982. "Memorial: Wallace Everette Pratt 1885–1981." *American Association of Petroleum Geologists Bulletin* vol. 66, no. 9: 1412.
Schrader, Frank Charles. *A Reconnaissance of Northern Alaska,* U.S. Geological Survey Prof. Paper 20. 1904: 139.
Simpson, John. 1855. "Observations upon western Esquimaux," *British Blue Book*, vol. 35: 917, 942.
Smith, P.S. and J.B. Mertie, Jr. 1930. "Geology and mineral resources of northwestern Alaska," U.S. Geological Survey Bulletin 815: 351.
Smith, R.J. 2002. *Human Events*: Online April 19, 2001: 4.
Smiley, Alfred Wilson. 1907. *A Few Scraps (Oily and Otherwise)*, Oil City, PA: The Derrick Publishing Company.
Specht, R.N., A.E. Brown, C.H. Selman and Carlisle. 1986. "Geophysical Case History, Prudhoe Bay Field," *Geophysics*, vol. 51: 1039–1049.
Standing, Thomas H. 2000. "Data shows steep Prudhoe Bay production decline," *Oil and Gas Journal*, vol. 98, Issue 40, October 2.
Stoney, George M., Lt. USN, "U.S. Naval Institute Proceedings." 1899: 810–823.
Sweet, John M. 1967–68. Personal diaries and expense account statements.

Thomas, Charles P., et al. 1991. "Energy: Wealth or Vanishing Opportunity," Alaska Oil Gas, Prepared for the U.S. Department of Energy, EG&G Idaho, Inc.
Trinity University. 2001. Various materials from their archives on E. de Leffingwell.

U.S. Geological Survey, Pacific Section AAPG, and Northern California Geological Society, 1970, Pacific Section, American Association of Petroleum Geologists.
U.S. Geological Survey, Map M, Alaska. 1988.

Van Dyke, Bill. 2000. "Prudhoe Bay production decline (Letter to Editor)," *Oil and Gas Journal*, vol. 98, Issue 43, October 23.

Wall, Bennett H. "Growth in a Changing Environment: A History of Standard Oil Company (New Jersey) 1950–1972 and Exxon Corporation 1972–1975."
Wilson, Leland E. 2000. Personal communication.
Woidneck, Keith, Philip Behrman, Charles Soule and Juliet Wu. 1987. "Reservoir Description of the Endicott Field, North Slope, Alaska." *Alaskan North Slope Geology,* Vol. One. The Pacific Section, Society of Economic Paleontologists and Mineralogists, and The Alaska Geological Society. www.worldstatesmen.org/Greenland.html

Yergin, Daniel. P. 571. *The Prise.* Yergin Free Press.

Index